McDougal Littell

Geometry

Concepts and Skills

Larson Boswell Stiff

CHAPTER 2 Resource Book

The Resource Book contains the wide variety of blackline masters available for Chapter 2. The blacklines are organized by lesson. Included are support materials for the teacher as well as practice, activities, applications, and assessment resources.

Contributing Authors

The authors wish to thank the following individuals for their contributions to the Chapter 2 Resource Book.

Rebecca Salmon Glus
Patrick M. Kelly
Lynn Hisae Lafferty
Cheryl A. Leech
Michelle H. McCarney
Jessica Pflueger
Barbara L. Power
Joanne V. Ricci

The authors wish to thank Meridian Creative Group for their contributions in creating this publication.

Copyright © 2003 by McDougal Littell Inc.
All rights reserved.

Permission is hereby granted to teachers to reprint or photocopy in classroom quantities the pages or sheets in this work that carry a McDougal Littell copyright notice. These pages are designed to be reproduced by teachers for use in their classes with accompanying McDougal Littell material, provided each copy made shows the copyright notice. Such copies may not be sold and further distribution is expressly prohibited. Except as authorized above, prior written permission must be obtained from McDougal Littell Inc. to reproduce or transmit this work or portions thereof in any other form or by any other electronic or mechanical means, including any information storage or retrieval system, unless expressly permitted by federal copyright laws. Address inquiries to Manager, Rights and Permissions, McDougal Littell Inc., P.O. Box 1667, Evanston, IL 60204.

ISBN: 0-618-14040-9

56789-BMW-04

Contents

Copyright © McDougal Littell Inc.
All rights reserved.

Contents

Copyright © McDougal Littell Inc.
All rights reserved.

Contents

Descriptions of Resources

This Chapter Resource Book is organized by lessons within the chapter in order to make your planning easier. The following materials are provided:

Tips for New Teachers These teaching notes provide both new and experienced teachers with useful teaching tips for each lesson, including tips about common errors and inclusion.

Parent Guide for Student Success This guide helps parents contribute to student success by providing an overview of the chapter along with questions and activities for parents and students to work on together.

Strategies for Reading Mathematics The first page teaches reading strategies to be applied to the current chapter and to later chapters. The second page is a visual glossary of key vocabulary.

Lesson Plans and Lesson Plans for Block Scheduling This planning template helps teachers select the materials they will use to teach each lesson from among the variety of materials available for the lesson. The block-scheduling version provides additional information about pacing.

Warm-Up Exercises and Daily Homework Quiz The warm-ups cover prerequisite skills that help prepare students for a given lesson. The quiz assesses students on the content of the previous lesson. (Transparencies also available)

Technology Activities and Keystrokes Keystrokes for the Geometry software and calculators are provided for each Technology Activity and Technology exercise in the Student Edition, along with additional Technology Activities to begin selected lessons.

Practice A and B These exercises offer additional practice for the material in each lesson, including application problems. There are two levels of practice for each lesson: A (basic) and B (average).

Reteaching with Practice These two pages provide additional instruction, worked-out examples, and practice exercises covering the key concepts and vocabulary in each lesson.

Quick Catch-Up for Absent Students This handy form makes it easy for teachers to let students who have been absent know what to do for homework and which activities or examples were covered in class.

Copyright © McDougal Littell Inc.
All rights reserved.

Contents

Learning Activities These enrichment activities apply the math taught in the lesson in an interesting way that lends itself to group work.

Real-Life Applications Students apply the mathematics covered in each lesson to solve an interesting real-life problem.

Quizzes The quizzes can be used to assess student progress on two or three lessons.

Brain Games Support These blackline masters make it easier for students to record their work on selected activities in the Student Edition.

Chapter Review Games and Activities This worksheet offers fun practice at the end of the chapter and provides an alternative way to review the chapter content in preparation for the Chapter Test.

Chapter Tests A and B These are tests that cover the most important skills taught in the chapter. There are two levels of test: A (basic) and B (average).

SAT/ACT Chapter Test This test also covers the most important skills taught in the chapter, but questions are in multiple-choice format. (See *Alternative Assessment* for multi-step problems.)

Alternative Assessment with Rubrics and Math Journal A journal exercise has students write about the mathematics in the chapter. A multi-step problem has students apply a variety of skills from the chapter and explain their reasoning. Solutions and a 4-point rubric are included.

Project with Rubric The project allows students to delve more deeply into a problem that applies the mathematics of the chapter. Teacher's notes and a 4-point rubric are included.

Cumulative Review These practice pages help students maintain skills from the current chapter and preceding chapters.

Cumulative Test This test covers the most important skills from the current chapter and preceding chapters. Cumulative Tests can be found in Chapters 3, 6, 9, and 11.

Copyright © McDougal Littell Inc.
All rights reserved.

CHAPTER 2

Tips for New Teachers

For use with Chapter 2

LESSON 2.1

TEACHING TIP When students learn a new term, they tend to try to understand the idea by focusing on the figure associated with the new term rather than focusing on the words used to define the term. This can lead to incorrect assumptions. For example, it may not be obvious to students that every segment has more than one bisector. The figure associated with the definition seems to imply that there is only one bisector. You should point out that the phrase "a segment bisector" implies that there can be more than one bisector.

TEACHING TIP Students need to recognize and use the algebraic form of the Midpoint Formula. They should also understand that the x-coordinate of the midpoint is the average of the x-coordinates of the endpoints of the segment and likewise for the y-coordinate of the midpoint.

LESSON 2.2

TEACHING TIP In Example 4 on page 63, point out that solving an equation for x is not the same as finding the measures of the angles. Checking the value of the variable by determining each angle measure is always important.

LESSON 2.3

INCLUSION Help students identify types of angles by using many visual examples drawn with a variety of orientations. Students with limited English proficiency can recognize the visual example and associate the description of the angle with it. If they have trouble when looking at angles printed on a page or on their paper, encourage them to turn their book or paper and view the angles from other directions or perspectives.

TEACHING TIP Theorems are introduced in this lesson and a numbering system is applied to them. Students should be reminded that like postulates, theorems are numbered to make their textbook reference easier. Stress that students must know the statement of the theorem and that using only the theorem number is not acceptable. Some theorems are named and you must decide when to have students write out the statement of the theorem rather than just using the name. Consider letting students use just the name once the chapter is completed and they fully understand the theorem.

LESSON 2.4

COMMON ERROR Students may think that vertical angles are adjacent when viewing a diagram. It may appear to them that a pair of opposite rays of the vertical angles is a common side because the rays form a line. Be sure they understand the difference between adjacent and vertical angles.

TEACHING TIP Before the students work on Exercises 38–50 on page 80, suggest that they redraw the figures with colored pencils or markers, using a different color for each line. This will help them to identify the vertical angles so that they can apply the Vertical Angles Theorem.

LESSON 2.5

INCLUSION It is often difficult for students to identify the hypothesis and conclusion of a statement that is not written explicitly as an if-then statement. You can suggest that the students first identify the conclusion by looking for the resulting action in the sentence, because the conclusion is the action that will take place when the hypothesis is satisfied. For instance, in Example 2 on page 82, the resulting action in part **a** is "is fun" and the resulting action in part **b** is "buying a CD." Then have the students identify the hypothesis by looking for the conditions that must be satisfied in order for the conclusion to occur. For instance, in Example 2**b,** have your students ask themselves "According to the statement, when will you buy the CD?"

Copyright © McDougal Littell Inc.
All rights reserved.

Tips for New Teachers

For use with Chapter 2

TEACHING TIP You should explain to your students what it means for an if-then statement to be true when discussing the Law of Detachment. In Example 3 on page 83, the statement is a known fact. In Example 4 on page 84, the if-then statement is true because it is a result of the Vertical Angles Theorem. You may want to give an example of a false if-then statement: "if $x^2 = 16$, then $x = 4$." This is false since x could be -4.

Lesson 2.6

TEACHING TIP Once students understand the Transitive Property of Equality, it may be used in a slightly different format than presented at the beginning of this lesson. This is evident in Example 3 on page 90, where $m\angle 1 + m\angle 3 = 180°$ and $m\angle 2 + m\angle 3 = 180°$ leads to a statement that $m\angle 1 + m\angle 3 = m\angle 2 + m\angle 3$ by the Transitive Property of Equality. Students need to be reassured that "if $a = b$ and $b = c$, then $a = c$" is equivalent to "if $a = b$ and $c = b$, then $a = c$." You could point out that changing $b = c$ to $c = b$ is justified by the Symmetric Property of Equality.

Outside Resources

BOOKS/PERIODICALS

Wilson, Patricia A. "Activities: Understanding Angles: Wedges to Degrees." *Mathematics Teacher* (April 1990); pp. 294–300.

May, Beverly A. " A Different Approach to Teaching the Midpoint Formula." *Mathematics Teacher* (May 1989); pp. 344–345.

ACTIVITIES/MANIPULATIVES

Piccioto, Henry. *Geometry Labs*. Collection of 78 blackline masters to use with manipulatives to promote reasoning skills. Emeryville, CA. Key Curriculum Press.

SOFTWARE

Geometry Activities for Middle School Students with Geometer's Sketchpad. Explore points, lines and angles, triangles, quadrilaterals, symmetry, and transformations.

Copyright © McDougal Littell Inc.
All rights reserved.

NAME ______________________ DATE __________

Parent Guide for Student Success

For use with Chapter 2

Chapter Overview One way that you can help your student to succeed in Chapter 2 is by discussing the lesson goals in the chart below. When a lesson is completed, you can ask your student the following questions. "What were the goals of the lesson? What new words and formulas did you learn? How can you apply the ideas of the lesson to your life?"

Lesson Title	*Lesson Goals*	*Key Applications*
2.1: Segment Bisectors	Bisect a segment. Find the coordinates of the midpoint of a segment.	• Biking • Strike Zone • Latitude-Longitude Coordinates
2.2: Angle Bisectors	Bisect an angle.	• Kites • Fans • Paper Airplanes • Lasers
2.3: Complementary and Supplementary Angles	Find measures of complementary and supplementary angles.	• Bridges • Beach Chairs
2.4: Vertical Angles	Find the measures of angles formed by intersecting lines.	• Flags • Drafting Table
2.5: If-Then Statements and Deductive Reasoning	Use if-then statements. Apply laws of logic.	• Advertising • Quotes of Wisdom
2.6: Properties of Equality and Congruence	Use properties of equality and congruence.	• Heights

Visual Strategy

Picturing Theorems is the visual strategy featured in Chapter 2 (see page 52). Be sure that your student keeps a list of theorems in his or her math notebook. To help remember a theorem, encourage your student to draw a sketch to go with the theorem that uses specific measures. Have your student tell you about each theorem the day it is introduced in class.

Copyright © McDougal Littell Inc.
All rights reserved.

CHAPTER 2 CONTINUED

Chapter Support

NAME ______________________________ DATE __________

Parent Guide for Student Success

For use with Chapter 2

Key Ideas **Your student can demonstrate understanding of key concepts by working through the following exercises with you.**

Lesson	Exercise
2.1	$\overline{PQ}$ has midpoint M. P has coordinates $(-3, 5)$. Q has coordinates $(7, -1)$. Find the coordinates of M.
2.2	$\overrightarrow{EG}$ bisects $\angle DEF$. Find $m\angle GEF$.
2.3	Find the measure of a complement and a supplement of a 36° angle.
2.4	Find $m\angle 1$, $m\angle 2$, and $m\angle 3$.
2.5	Use the Law of Syllogism to write the statement that follows from the pair of true statements. If Roger earns money, he buys new clothes. If Roger works, he earns money.
2.6	Use the Transitive Property of Equality to complete the statement. If $m\angle K = 90° - m\angle J$ and $90° - m\angle J = m\angle M$, then ________.

Home Involvement Activity

Directions: Talk to a number of different people and develop a list of folk beliefs. For example, some people believe that it is going to rain when a person's knees hurt. State each belief as an if-then statement. Are the if-then statements true? Are there any examples of the Law of Syllogism in your list?

Answers

2.1: (2, 2) **2.2:** 55° **2.3:** 54°, 144° **2.4:** 135°, 45°, 135° **2.5:** If Roger works, then he buys new clothes. **2.6:** $m\angle K = m\angle M$

Copyright © McDougal Littell Inc.
All rights reserved.

CHAPTER 2

NAME _______________________________ DATE ________

Strategies for Reading Mathematics

For use with Chapter 2

Strategy: Reading Vocabulary and Taking Notes in Geometry

In Algebra, you learned many algebraic terms. Now that you are taking geometry, you will need to understand many geometric terms. Important vocabulary terms you need to learn are printed in **heavy type like this**. Use diagrams to give examples of what these terms mean in your notebook. The notebook below shows a sample of the notes you might take about the vocabulary terms in the next paragraph.

Two angles are **adjacent** if they share a common vertex and side but have no common interior points. Two adjacent angles are a **linear pair** if their noncommon sides are opposite rays. Two angles are **vertical angles** if their sides form two pairs of opposite rays.

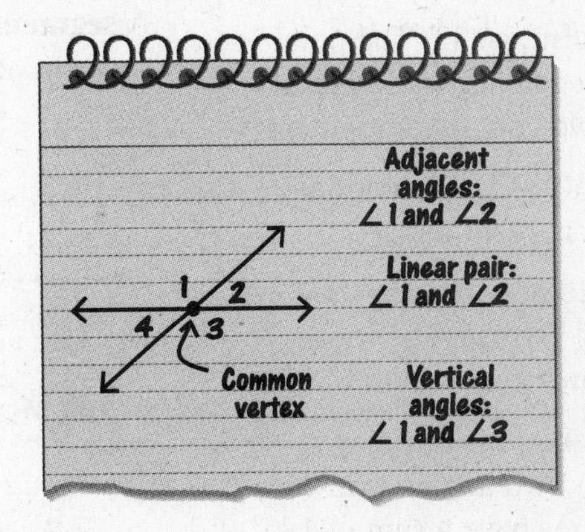

STUDY TIP
Reading Vocabulary

Look for terms printed in heavy type. Read the sentence that contains the term to study its definition. Then read the entire paragraph containing the sentence to be sure you understand the entire definition.

STUDY TIP
Taking Notes

Take brief but complete notes. Write key words from the definition of each term to help you remember the meaning of the term. Draw labeled pictures or diagrams to illustrate the term. You can also include brief examples when appropriate.

Questions

1. Name other angles shown in the notebook that are adjacent. Explain how you know they are adjacent.
2. Name other angles shown in the notebook that are linear pairs. Explain how you know they are linear pairs.
3. Name other angles shown in the notebook that are vertical angles. Explain how you know they are vertical angles.
4. Which angles in the figure at the right are vertical angles? How can you use the drawing in the notebook to help you decide?

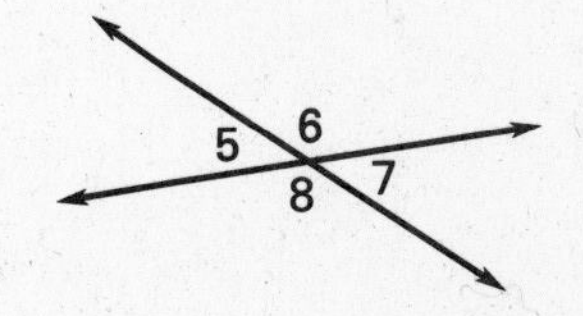

Copyright © McDougal Littell Inc.
All rights reserved.

NAME ______________________ DATE __________

Strategies for Reading Mathematics

For use with Chapter 2

Visual Glossary

The Study Guide on page 52 lists the key words for Chapter 2. Use the visual glossary below to help you understand some of the key words in Chapter 2. You may want to copy the diagrams into your notebook and refer to them as you complete the chapter.

GLOSSARY

midpoint (p. 53) The point on a segment that divides it into two congruent segments.

bisect (p. 53) To divide into two congruent parts.

complementary angles (p. 67) Two angles whose measures have a sum of 90°.

supplementary angles (p. 67) Two angles whose measures have a sum of 180°.

adjacent angles (p. 68) Two angles with a common vertex and side but no common interior points.

theorem (p. 69) A true statement that follows from other true statements.

vertical angles (p. 75) Two angles that are not adjacent and whose sides are formed by two intersecting lines.

linear pair (p. 75) Two adjacent angles whose noncommon sides are on the same line.

Segment and Angle Bisectors

A **segment bisector** is a segment, line, ray, or plane that intersects a segment at its midpoint.

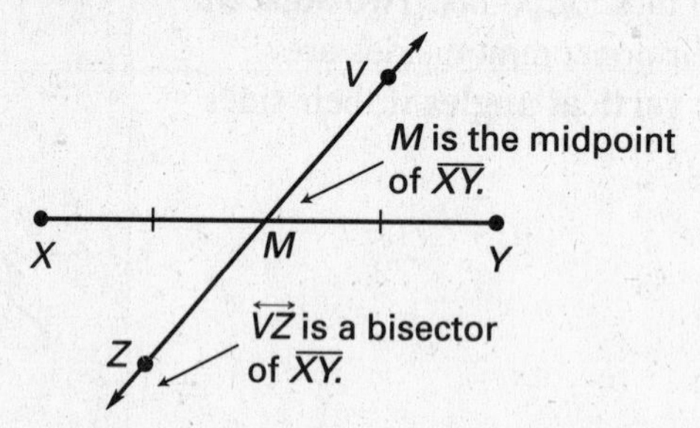

An **angle bisector** is a ray that divides an angle into two adjacent angles that are congruent.

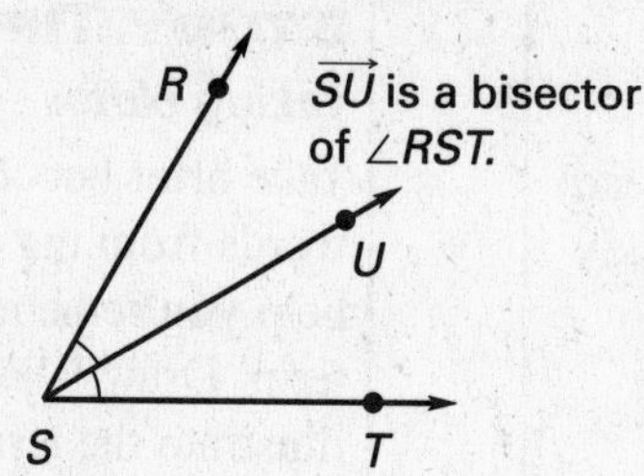

Complementary and Supplementary Angles

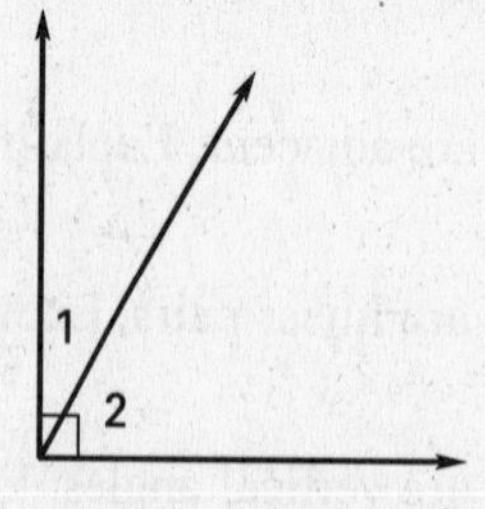

$m\angle 1 + m\angle 2 = 90°$, so $\angle 1$ and $\angle 2$ are complements.

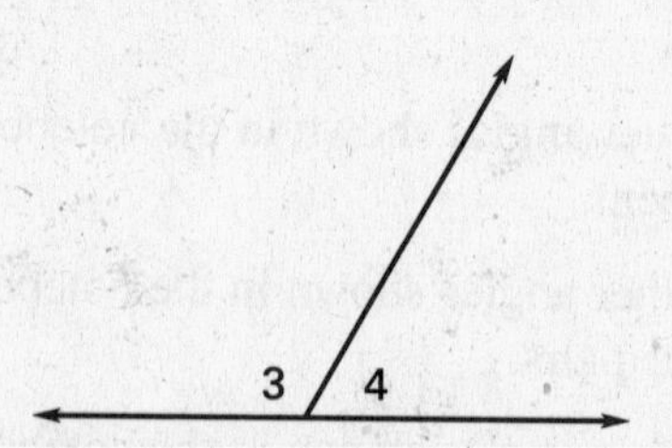

$m\angle 3 + m\angle 4 = 180°$, so $\angle 3$ and $\angle 4$ are supplements.

Copyright © McDougal Littell Inc.
All rights reserved.

TEACHER'S NAME ______________________ CLASS _______ ROOM _______ DATE _______

Lesson Plan

2-day lesson (See *Pacing the Chapter,* TE page 50A)

For use with pages 53–59

GOAL **Bisect a segment. Find the coordinates of the midpoint of a segment.**

State/Local Objectives ______________________________

____

✓ Check the items you wish to use for this lesson.

STARTING OPTIONS

____ Strategies for Reading Mathematics: CRB pages 5–6
____ Warm-Up: CRB page 9 or Transparencies

TEACHING OPTIONS

____ Geo-Activity: SE page 53
____ Examples: Day 1: 1–2, SE page 54; Day 2: 3–4, SE pages 54–55
____ Extra Examples: TE pages 54–55
____ Checkpoint Exercises: Day 1: 1–2, SE page 54; Day 2: 3–5, SE page 55
____ Concept Check: TE page 55
____ Guided Practice Exercises: Day 1: 1–5, SE page 56; Day 2: 6–10, SE page 56

APPLY/HOMEWORK

Homework Assignment

____ Basic: Day 1: pp. 56–59 Exs. 11–24, 46–51, 53–63 odd
Day 2: SRH p. 672 Exs. 1–19 odd; pp. 56–59 Exs. 26–32, 38, 44, 45, 52–62 even
____ Average: Day 1: pp. 56–59 Exs. 11–19 odd, 20–23, 25, 42, 46–57
Day 2: pp. 56–59 Exs. 26–40, 44, 45, 58–63
____ Advanced: Day 1: pp. 56–59 Exs. 13–17, 20–23, 25, 42, 46–62 even
Day 2: pp. 56–59 Exs. 24–34 even, 36–41, 43*, 44, 45; EC: classzone.com

Reteaching the Lesson

____ Practice Masters: CRB pages 10–11 (Level A, Level B)
____ Reteaching with Practice: CRB pages 12–13 or Practice Workbook with Examples; Resources in Spanish

Extending the Lesson

____ Learning Activity: CRB page 15
____ Challenge: SE page 59; classzone.com

ASSESSMENT OPTIONS

____ Daily Quiz (2.1): TE page 59, CRB page 18, or Transparencies
____ Standardized Test Practice: SE page 59; Transparencies

Notes ______________________________

____

____

Copyright © McDougal Littell Inc.
All rights reserved.

TEACHER'S NAME ______________________ CLASS ________ ROOM ________ DATE ________

Lesson Plan for Block Scheduling

1-block lesson (See *Pacing the Chapter,* TE page 50A) For use with pages 53–59

GOAL **Bisect a segment. Find the coordinates of the midpoint of a segment.**

State/Local Objectives ______________________

CHAPTER PACING GUIDE	
Day	**Lesson**
1	**2.1**
2	2.2
3	2.3
4	2.4
5	2.5
6	2.6
7	Ch. 2 Review and Assess

✓ Check the items you wish to use for this lesson.

STARTING OPTIONS

____ Strategies for Reading Mathematics: CRB pages 5–6
____ Homework Quiz (1.6): TE page 41, CRB page 9, or Transparencies
____ Warm–Up: CRB page 9 or Transparencies

TEACHING OPTIONS

____ Geo-Activity: SE page 53
____ Examples: 1–4, SE pages 54–55
____ Extra Examples: TE pages 54–55
____ Checkpoint Exercises: 1–5, SE pages 54–55
____ Concept Check: TE page 55
____ Guided Practice Exercises: 1–10, SE page 56

APPLY/HOMEWORK

Homework Assignment

____ Block Schedule: pp. 56–59 Exs. 11–19 odd, 20–40, 44–63

Reteaching the Lesson

____ Practice Masters: CRB pages 10–11 (Level A, Level B)
____ Reteaching with Practice: CRB pages 12–13 or Practice Workbook with Examples; Resources in Spanish

Extending the Lesson

____ Learning Activity: CRB page 15
____ Challenge: SE page 59; classzone.com

ASSESSMENT OPTIONS

____ Daily Quiz (2.1): TE page 59, CRB page 18, or Transparencies
____ Standardized Test Practice: SE page 59; Transparencies

Notes ______________________

Copyright © McDougal Littell Inc.
All rights reserved.

NAME ______________________ DATE ________

Available as a transparency

WARM-UP EXERCISES

For use before Lesson 2.1, pages 53–59

Find the value of the variable.

1. $7z = 49$
2. $2x - 4 = 18$
3. $y + 5 = 27$

Find the coordinate of the point if $x = 2$ and $y = -1$.

4. $\left(\frac{x+2}{2}, \frac{y-1}{2}\right)$

5. $\left(\frac{x-4}{2}, \frac{y+5}{2}\right)$

DAILY HOMEWORK QUIZ

For use after Lesson 1.6, pages 34–41

Use the figure at the right.

1. Use a protractor to approximate the measure of $\angle ABC$.
2. Use your answer to Exercise 1 and the Angle Addition Postulate to find the measure of $\angle ABD$.
3. Classify each of the three angles shown in the figure as *acute*, *right*, *obtuse*, or *straight*.

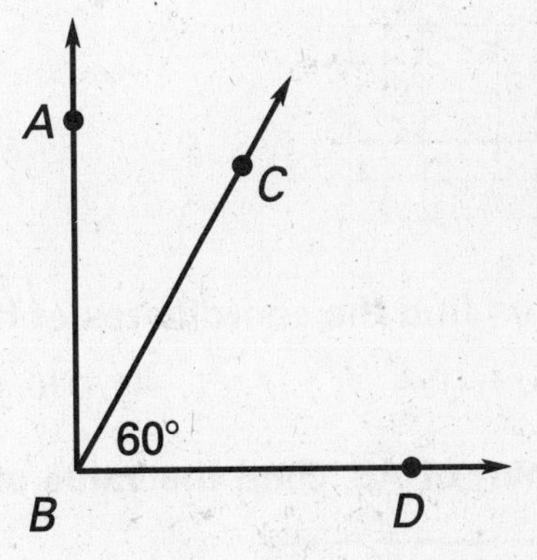

Copyright © McDougal Littell Inc.
All rights reserved.

NAME ______________________ DATE __________

Practice A

For use with pages 53–59

Complete the statement.

1. The __?__ of a segment is the point on the segment that divides it into two congruent segments.
2. A __?__ is a segment, ray, line, or plane that intersects a segment at its midpoint.
3. To __?__ a segment means to divide the segment into two congruent segments.

***M* is the midpoint of the segment. Find the segment lengths.**

4. Find *TM* and *MR*.

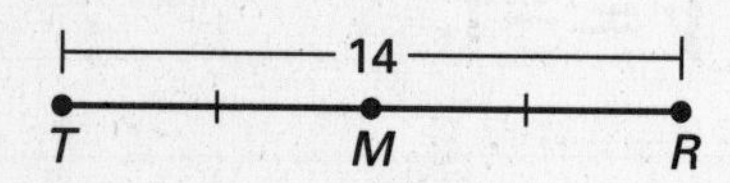

5. Find *FM* and *MD*.

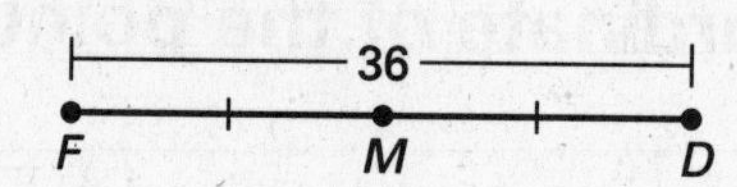

6. Find *MR* and *QR*.

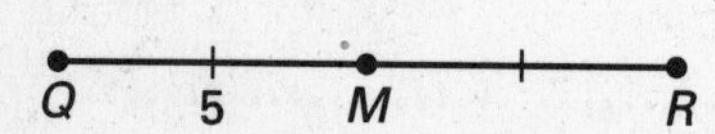

7. Find *KM* and *KL*.

Use the Midpoint Formula to find the coordinates of the midpoint of $\overline{CD}$.

8.

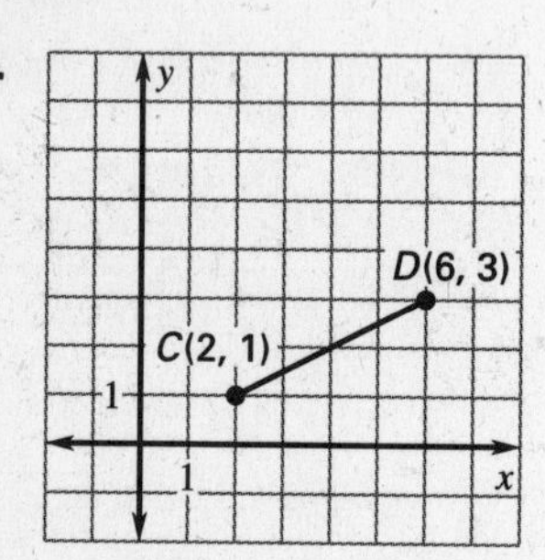

9.

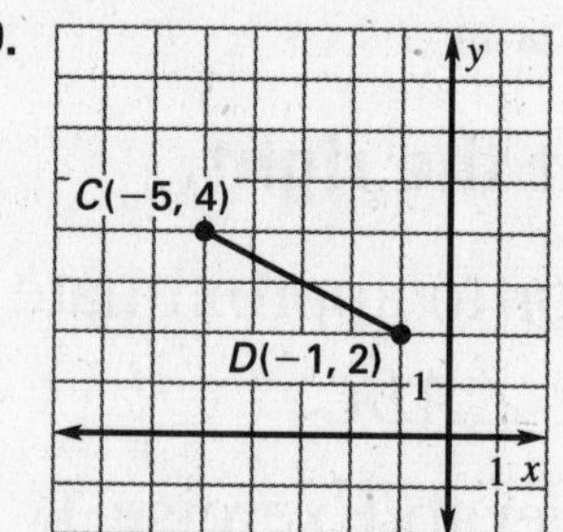

Sketch $\overline{PQ}$. Then find the coordinates of its midpoint.

10. $P(0, 0)$, $Q(6, -4)$

11. $P(0, 8)$, $Q(2, 6)$

12. $P(1, 2)$, $Q(-5, 0)$

***M* is the midpoint of $\overline{AB}$. Find the value of *x*.**

13. A ——— $2x$ ——— M ——— 18 ——— B

14. A ——— $3x$ ——— M ——— 24 ——— B

15. A balance beam is shown at the right. Your gymnastics routine includes a jump at the midpoint *M* of the beam. If the length of the beam is 500 centimeters, what is the distance from the end of the beam to the jump location?

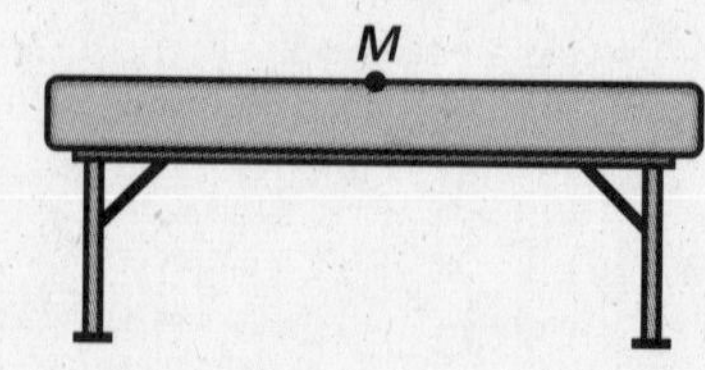

Copyright © McDougal Littell Inc.
All rights reserved.

LESSON 2.1

NAME ______________________ DATE ______

Practice B

For use with pages 53–59

Determine whether M is the midpoint of $\overline{JK}$. Explain your reasoning.

1. 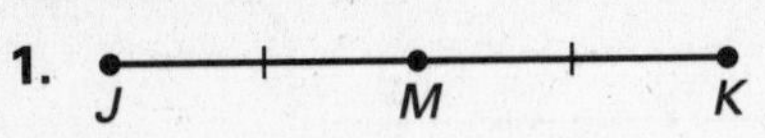

2.  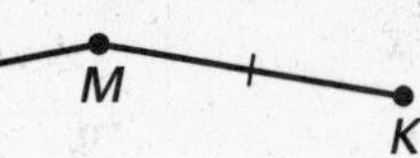

3. 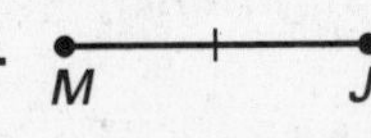 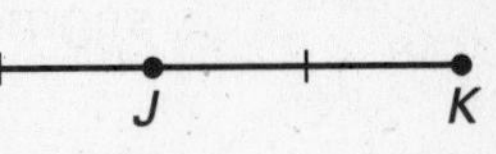

Find the segment lengths, given that M is the midpoint of the segment.

4. Find PM and MQ.

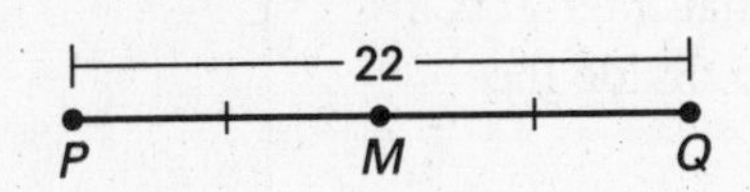

5. Find TM and MS.

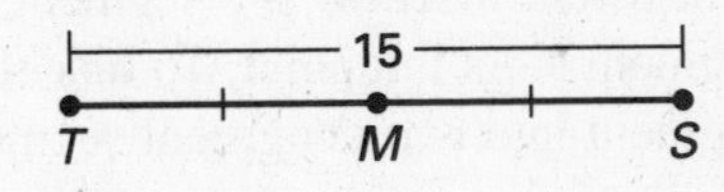

Line ℓ bisects the segment. Find the segment lengths.

6. Find AN and AB.

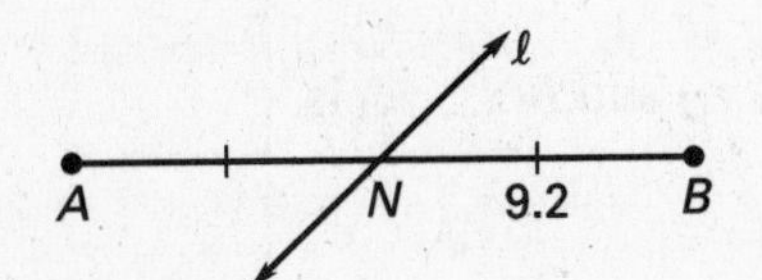

7. Find FG and EG.

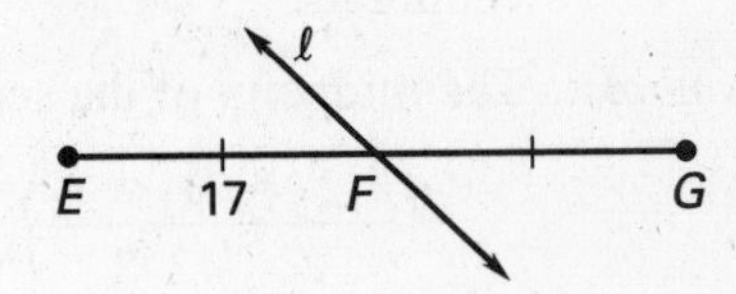

Line ℓ bisects the segment. Find the value of x.

8. 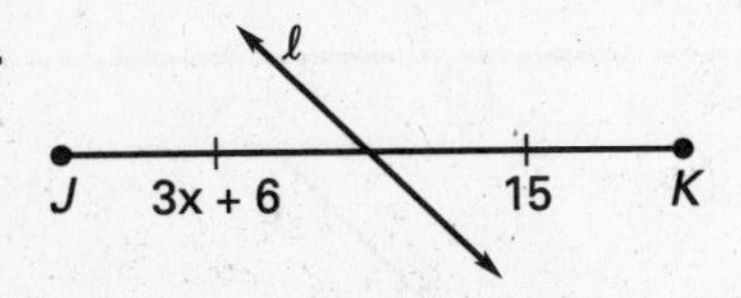

9. 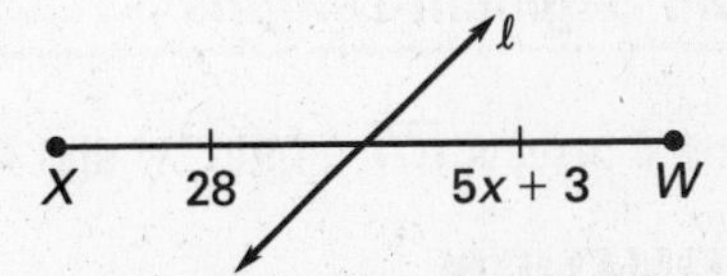

Find the coordinates of the midpoint of $\overline{FG}$.

10. $F(-2, 3), G(4, -1)$

11. $F(1, -5), G(4, -3)$

12. $F(-6, 1), G(-2, 7)$

13. $F(3, -2), G(4, -4)$

14. $F(-1, 7), G(5, -2)$

15. $F(-3, 6), G(-1, 2)$

16. You are planting flowers in a border that is 11 feet long. If you want to plant a large rosebush halfway along the border, how far is the bush from the ends of the border?

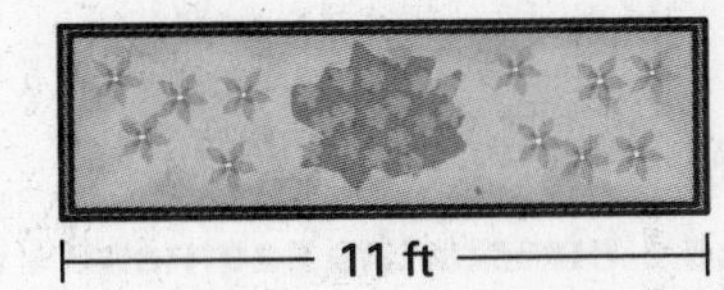

11 ft

Copyright © McDougal Littell Inc.
All rights reserved.

LESSON 2.1

NAME ______________________________ DATE __________

Reteaching with Practice

For use with pages 53–59

GOAL **Bisect a segment. Find the coordinates of the midpoint of a segment.**

VOCABULARY

The **midpoint** of a segment is the point on the segment that divides it into two congruent segments.

A **segment bisector** is a segment, ray, line, or plane that intersects a segment at its midpoint. To **bisect** a segment means to divide the segment into two congruent segments.

The Midpoint Formula:

Words: The coordinates of the midpoint of a segment are the averages of the x-coordinates and the y-coordinates of the endpoints.

Symbols: The midpoint of the segment joining $A(x_1, y_1)$ and $B(x_2, y_2)$ is

$$\left(\frac{x_1 + x_2}{2}, \frac{y_1 + y_2}{2}\right).$$

EXAMPLE 1 *Find Segment Lengths*

Line ℓ bisects $\overline{CD}$. Find CM and CD.

C M 12 D ℓ

SOLUTION

M is the midpoint of $\overline{CD}$, so $CM = MD$. Therefore, $CM = 12$.

You know that CD is twice the length of $\overline{MD}$.

$$CD = 2 \cdot MD = 2 \cdot 12 = 24$$

So, $CM = 12$ and $CD = 24$.

Exercises for Example 1

Line ℓ bisects the segment. Find the value of x.

1.
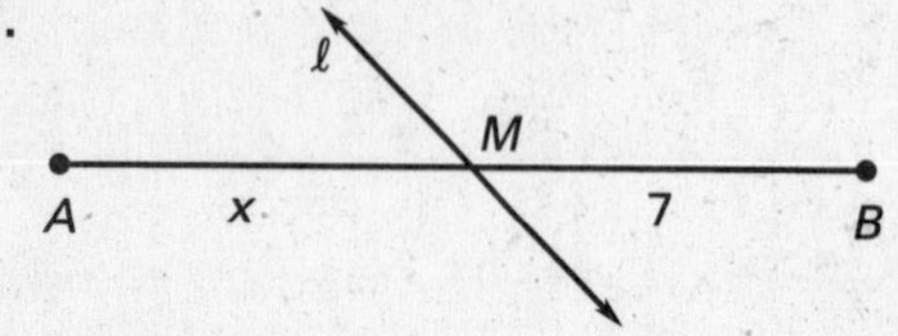

2.
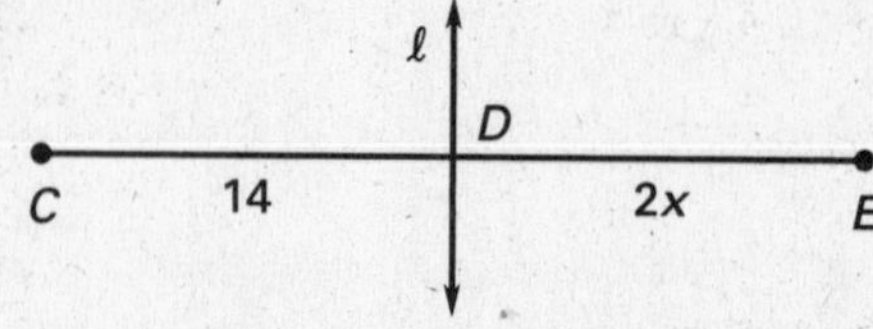

3.
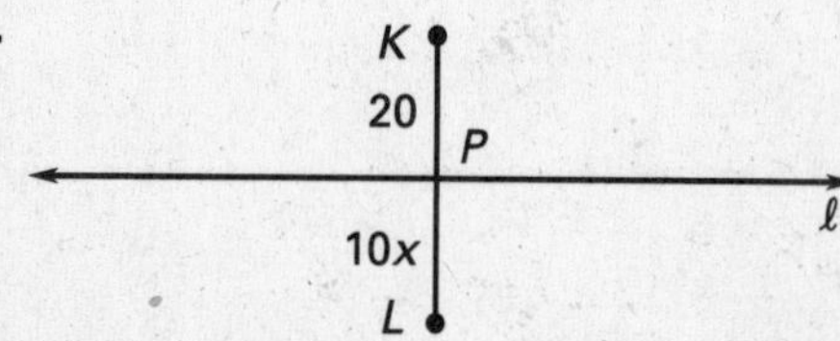

4.
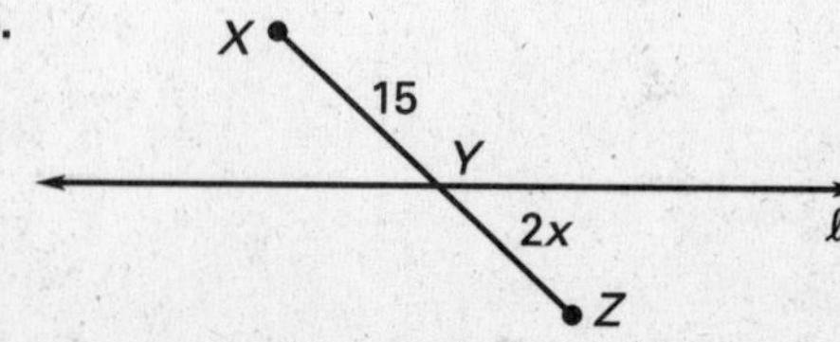

Copyright © McDougal Littell Inc.
All rights reserved.

LESSON 2.1 CONTINUED

NAME _______________ DATE _______

Reteaching with Practice

For use with pages 53–59

EXAMPLE 2 Use Algebra with Segment Lengths

M is the midpoint of $\overline{CD}$.
Find the value of x.

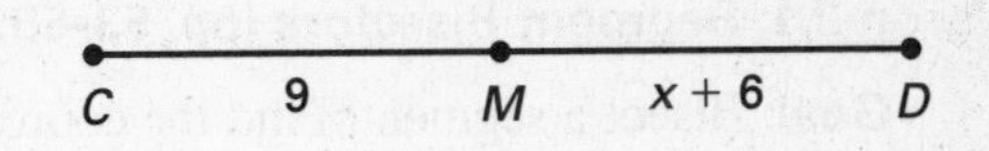

SOLUTION

$CM = MD$	M is the midpoint of $\overline{CD}$.
$9 = x + 6$	Substitute 9 for CM and $x + 6$ for MD.
$9 - 6 = x + 6 - 6$	Subtract 6 from each side.
$3 = x$	Simplify.

Exercises for Example 2

M is the midpoint of the segment. Find the segment lengths.

5. Find AM and MB.

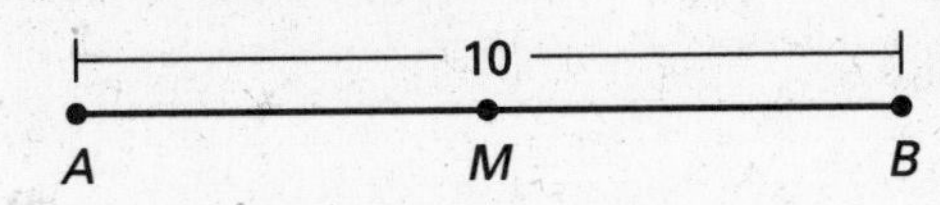

6. Find ML and KL.

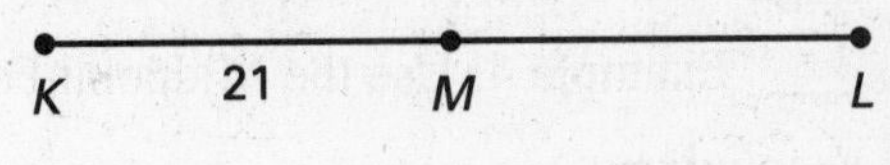

7. Find ST and SM.

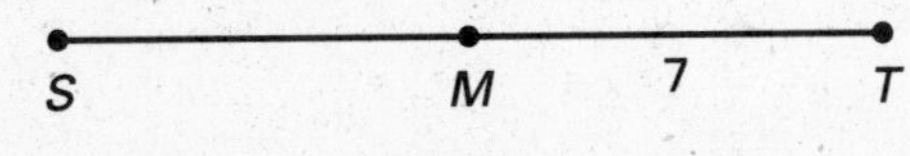

8. Find CM and MD.

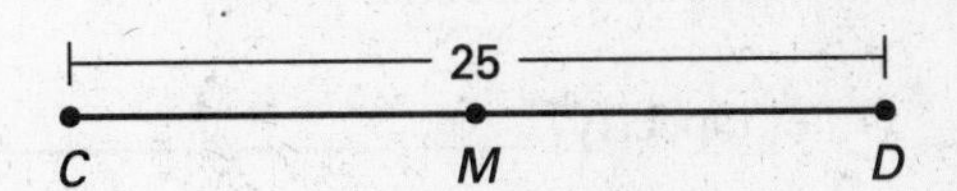

EXAMPLE 3 Use the Midpoint Formula

Find the coordinates of the midpoint of $\overline{CD}$.

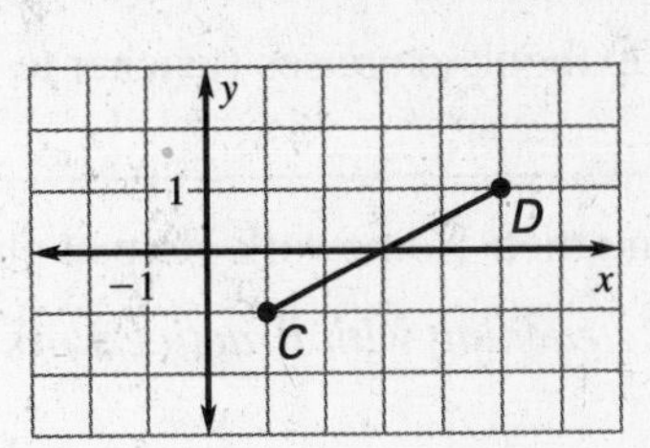

SOLUTION

Let $(x_1, y_1) = (1, -1)$ and $(x_2, y_2) = (5, 1)$.

$$M = \left(\frac{x_1 + x_2}{2}, \frac{y_1 + y_2}{2}\right) = \left(\frac{1 + 5}{2}, \frac{-1 + 1}{2}\right) = (3, 0)$$

Exercises for Example 3

Sketch $\overline{CD}$. Then find the coordinates of its midpoint.

9. $C(2, 2), D(4, 8)$ **10.** $C(-5, 4), D(1, -2)$ **11.** $C(2, -3), D(3, -7)$

Copyright © McDougal Littell Inc.
All rights reserved.

LESSON 2.1

NAME ______________________ DATE ________

Quick Catch-Up for Absent Students

For use with pages 53–59

The items checked below were covered in class on (date missed) ________

Lesson 2.1: Segment Bisectors (pp. 53–55)

____ **Goal:** Bisect a segment. Find the coordinates of the midpoint of a segment.

Material Covered:

____ Geo-Activity: Folding a Segment Bisector

____ Student Help: Vocabulary Tip

____ Student Help: Study Tip

____ Example 1: Find Segment Lengths

____ Example 2: Find Segment Lengths

____ Example 3: Use Algebra with Segment Lengths

____ Student Help: Reading Tip

____ Student Help: Skills Review

____ Example 4: Use the Midpoint Formula

Vocabulary:

midpoint, p. 53 | bisect, p. 53

segment bisector, p. 53

____ Other (specify) ______________________

____

Homework and Additional Learning Support

____ Textbook exercises (teacher to specify) pp. 56–59

____

____ Internet: Homework Help at classzone.com

____ *Reteaching with Practice* worksheet

Lesson 2.1

Copyright © McDougal Littell Inc.
All rights reserved.

NAME ________________________ DATE ________

Learning Activity

For use with pages 53–59

GOAL **Use the properties of segment congruence to understand why an inch is divided into 16 equal parts.**

Materials: paper, pencil, ruler

Exploring Statements about Segments

Every student has used a ruler to measure distances. In this activity, your group will work with line segments to better understand why an inch is divided into 16 equal parts.

Instructions

1. Have one member of your group draw a line segment that is exactly 8 inches long. It is important that this measurement be as precise as possible. Label the endpoints of this segment A and B.

   A ———————— B

2. The next member of the group should bisect the segment by folding, as shown on page 53 of the textbook. The 8-inch segment should now be divided into two line segments that are equal in length. Label the midpoint C. Measure $\overline{AC}$ and $\overline{CB}$ to confirm that they are equal in length.

   A ———— C ———— B

3. Have another group member repeat the process for one of the two equal line segments.

4. Continue with each member of the group repeating this process until the original 8-inch segment has been divided into sixteen equal parts.

5. Finally, have a group member measure the sixteen line segments to confirm that they are equal in length.

Analyzing the Results

1. What is the length of $\overline{AC}$? What is the length of $\overline{CB}$?

2. Write an equation that relates the length of $\overline{AB}$ to the length of $\overline{AC}$.

3. Explain why the customary measuring system divides inches into 2, 4, 8, and 16 equal parts.

Copyright © McDougal Littell Inc.
All rights reserved.

LESSON 2.2

TEACHER'S NAME ______________________ CLASS ______ ROOM ______ DATE ______

Lesson Plan

2-day lesson (See *Pacing the Chapter,* TE page 50A)

For use with pages 60–66

GOAL **Bisect an angle.**

State/Local Objectives ____

____

✓ Check the items you wish to use for this lesson.

STARTING OPTIONS

____ Homework Check (2.1): TE page 56; Answer Transparencies
____ Homework Quiz (2.1): TE page 59, CRB page 18, or Transparencies
____ Warm-Up: CRB page 18 or Transparencies

TEACHING OPTIONS

____ Activity: SE page 60
____ Examples: Day 1: 1–2, SE pages 61–62; Day 2: 3–4, SE pages 62–63
____ Extra Examples: TE pages 62–63; Internet Help at classzone.com
____ Checkpoint Exercises: Day 1: 1–6, SE pages 61–62; Day 2: 7–10, SE page 63
____ Technology Keystrokes for Ex. 31 on SE page 66: CRB page 19
____ Concept Check: TE page 63
____ Guided Practice Exercises: Day 1: 1–7, SE page 64
____ Visualize It! Transparencies: 6

APPLY/HOMEWORK

Homework Assignment

____ Basic: Day 1: EP p. 676 Exs. 29–31; pp. 64–66 Exs. 8–22, 33, 42–47
Day 2: SRH p. 673 Exs. 4–9, 19–21; pp. 64–66 Exs. 24–26, 28–30, 34–41
____ Average: Day 1: pp. 64–66 Exs. 9–23 odd, 33–47 odd
Day 2: pp. 64–66 Exs. 24–31, 34–46 even
____ Advanced: Day 1: pp. 64–66 Exs. 9–23 odd, 32*, 33–47 odd
Day 2: pp. 64–66 Exs. 24–31, 34–46 even; EC: classzone.com

Reteaching the Lesson

____ Practice Masters: CRB pages 20–21 (Level A, Level B)
____ Reteaching with Practice: CRB pages 22–23 or Practice Workbook with Examples; Resources in Spanish

Extending the Lesson

____ Real-Life Application: CRB page 25
____ Challenge: SE page 66; classzone.com

ASSESSMENT OPTIONS

____ Daily Quiz (2.2): TE page 66, CRB page 28, or Transparencies
____ Standardized Test Practice: SE page 66; Transparencies

Notes ____

____

____

Copyright © McDougal Littell Inc.
All rights reserved.

LESSON 2.2

TEACHER'S NAME ______________ CLASS ______ ROOM ______ DATE ______

Lesson Plan for Block Scheduling

1-block lesson (See *Pacing the Chapter,* TE page 50A)

For use with pages 60–66

GOAL Bisect an angle.

State/Local Objectives ______________________________

CHAPTER PACING GUIDE	
Day	Lesson
1	2.1
2	**2.2**
3	2.3
4	2.4
5	2.5
6	2.6
7	Ch. 2 Review and Assess

✓ **Check the items you wish to use for this lesson.**

STARTING OPTIONS

____ Homework Check (2.1): TE page 56; Answer Transparencies
____ Homework Quiz (2.1): TE page 59, CRB page 18, or Transparencies
____ Warm-Up: CRB page 18 or Transparencies

TEACHING OPTIONS

____ Activity: SE page 60
____ Examples: 1–4, SE pages 61–63
____ Extra Examples: TE pages 62–63; Internet Help at classzone.com
____ Checkpoint Exercises: 1–10, SE pages 61–63
____ Technology Keystrokes for Ex. 31 on SE page 66: CRB page 19
____ Concept Check: TE page 63
____ Guided Practice Exercises: 1–7, SE page 64
____ Visualize It! Transparencies: 6

APPLY/HOMEWORK

Homework Assignment

____ Block Schedule: pp. 64–66 Exs. 9–23 odd, 24–31, 33–47

Reteaching the Lesson

____ Practice Masters: CRB pages 20–21 (Level A, Level B)
____ Reteaching with Practice: CRB pages 22–23 or Practice Workbook with Examples; Resources in Spanish

Extending the Lesson

____ Real-Life Application: CRB page 25
____ Challenge: SE page 66; classzone.com

ASSESSMENT OPTIONS

____ Daily Quiz: (2.2): TE page 66, CRB page 28, or Transparencies
____ Standardized Test Practice: SE page 66; Transparencies

Notes ______________________________

Copyright © McDougal Littell Inc.
All rights reserved.

LESSON 2.2

NAME ______________________ DATE __________

Available as a transparency

WARM-UP EXERCISES

For use before Lesson 2.2, pages 60–66

Find the measure of the angle.

1. $m\angle A = 2 \cdot 58°$

2. $m\angle B = \frac{1}{2} \cdot 130°$

3. $\frac{1}{2}m\angle C = 26°$

Find the value of *x*.

4. $2x - 9 = 9$

5. $8x = 7x + 15$

DAILY HOMEWORK QUIZ

For use after Lesson 2.1, pages 53–59

1. Line ℓ bisects $\overline{FH}$. Find GH and FH.

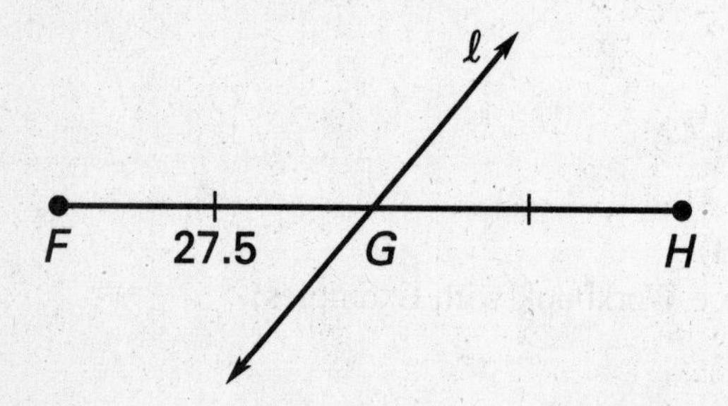

2. M is the midpoint of $\overline{AB}$. Find the value of x.

A 7x − 2 M 33 B

3. Find the coordinates of the midpoint of the segment whose endpoints are $P(-2, -3)$ and $Q(2, 7)$.

4. Julia is hiking a trail that is 3.4 miles long. She plans to stop for lunch when she reaches the halfway point on the trail. How far will she hike before stopping to eat her lunch?

Copyright © McDougal Littell Inc.
All rights reserved.

NAME ______________________ DATE __________

Technology Keystrokes

For use with Exercise 31, page 66

Keystrokes for Exercise 31

TI-92

1. Draw triangle *ABC*.

   F3 3 (Move cursor to desired location.) ENTER *A*

   (Move cursor to desired location.) ENTER *B*

   (Move cursor to desired location.) ENTER *C*

2. Draw the angle bisector of $\angle BAC$.

   F4 5 (Move cursor to point *B*.) ENTER

   (Move cursor to point *A*.) ENTER

   (Move cursor to point *C*.) ENTER

3. Draw the midpoint of $\overline{BC}$.

   F4 3 (Move cursor to segment $\overline{BC}$.) ENTER *D*

4. Draw any of the vertices of triangle *ABC*.

   F1 1 (Move cursor to desired point.) Use the drag key [drag key] and the cursor pad to drag the point.

SKETCHPAD

1. Draw triangle *ABC*.
   Choose the segment tool. Draw $\overline{AB}$, $\overline{BC}$, and $\overline{CA}$ to form triangle *ABC*.
2. Construct the angle bisector of $\angle BAC$.
   Choose the selection arrow tool. Select point *B*, hold down the shift key and select points *A* and *C*.
   Choose **Angle Bisector** from the **Construct** menu.
3. Find the midpoint of $\overline{BC}$.
   Select $\overline{BC}$. Choose **Point at Midpoint** from the **Construct** menu.
4. Drag the points.
   Select any of the vertices of triangle *ABC* and drag.

Copyright © McDougal Littell Inc.
All rights reserved.

LESSON 2.2

NAME ______________________ DATE __________

Practice A

For use with pages 60–66

Complete the statement for the diagram at the right.

1. __?__ is bisected by __?__.
2. The measure of $\angle ABC$ is __?__ the measure of $\angle ABD$.

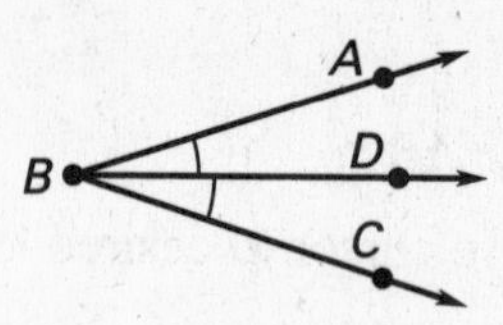

$\overrightarrow{GE}$ bisects $\angle FGH$. Find the angle measure.

3. Find $m\angle EGH$. 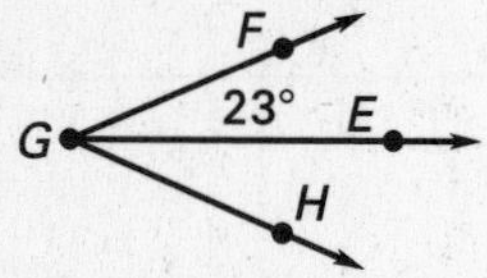
4. Find $m\angle FGE$.
5. Find $m\angle EGF$.

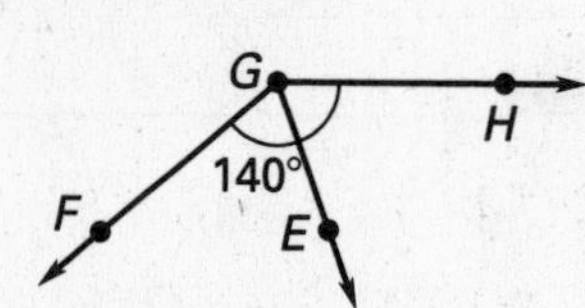

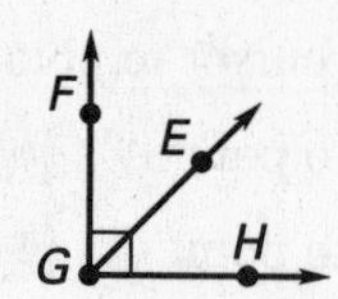

$\overrightarrow{BA}$ bisects $\angle DBC$. Find $m\angle CBA$ and $m\angle DBC$.

6.

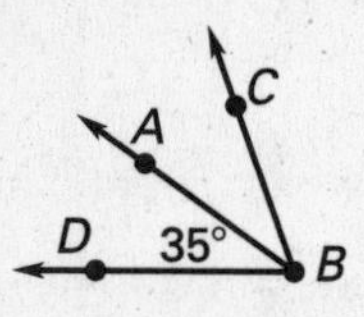

7.

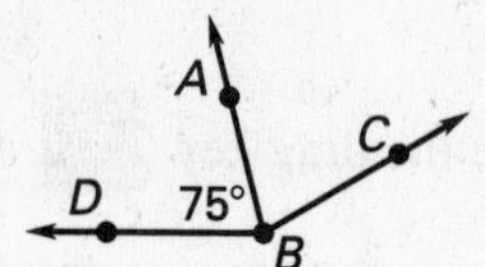

8.

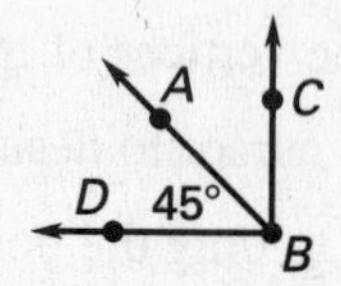

$\overrightarrow{JK}$ bisects $\angle GJH$. Find the value of x.

9.

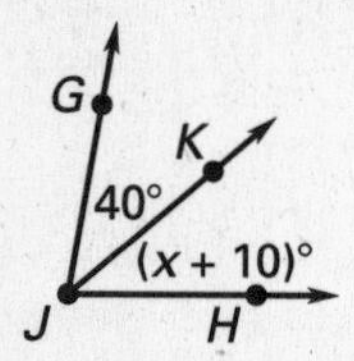

10.

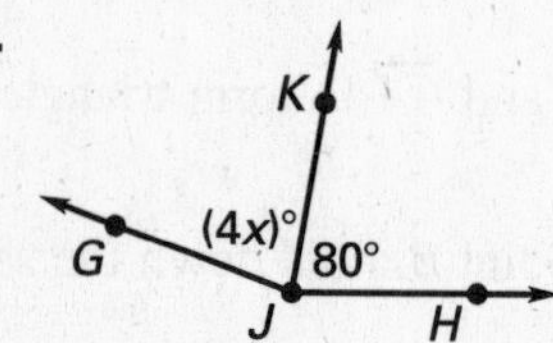

11.

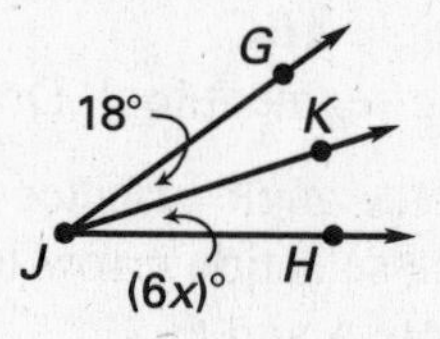

Use the diagram at the right. Decide whether the statement is *true* or *false*.

12. If $\overrightarrow{BD}$ bisects $\angle ABC$, then $\angle ABD \cong \angle ABC$.
13. If $\overrightarrow{BD}$ bisects $\angle ABC$, then $\angle DBC \cong \angle ABD$.
14. If $\overrightarrow{BD}$ bisects $\angle ABC$ and $m\angle ABD = 55°$, then $m\angle DBC = 55°$.
15. If $\overrightarrow{BD}$ bisects $\angle ABC$ and $m\angle ABC = 112°$, then $m\angle ABD = 61°$.

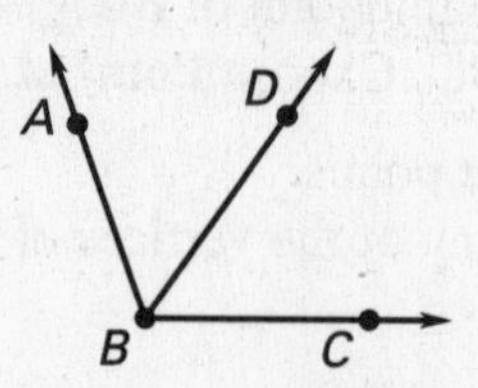

16. In the pup tent shown at the right, the two sides meet at the top to form a 72° angle. If the tent pole bisects the angle, what angle does the tent pole make with each of the sides?

Lesson 2.2

Copyright © McDougal Littell Inc.
All rights reserved.

LESSON 2.2

NAME ______________________________ DATE __________

Practice B

For use with pages 60–66

$\overrightarrow{QS}$ bisects $\angle PQR$. Find the angle measure.

1. Find $m\angle SQR$.

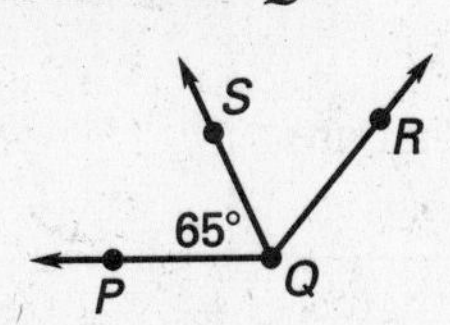

2. Find $m\angle PQS$.

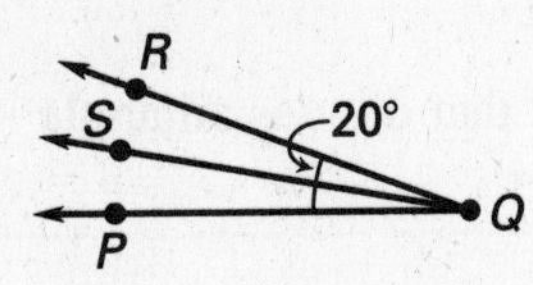

3. Find $m\angle RQS$.

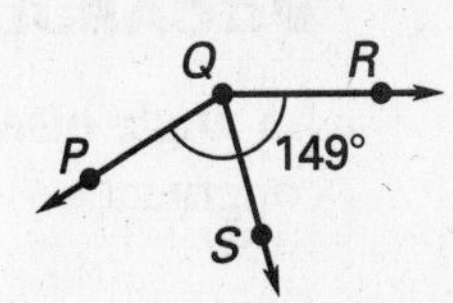

Find the measures of $\angle EFJ$ and $\angle JFG$.

4.

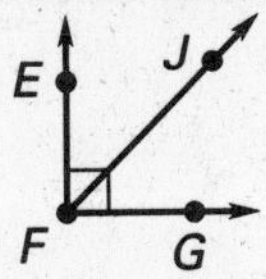

5.

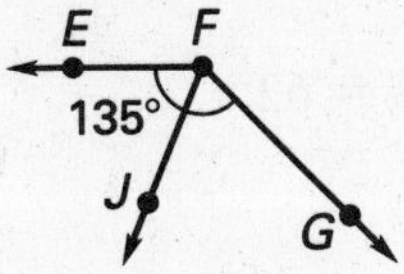

6.

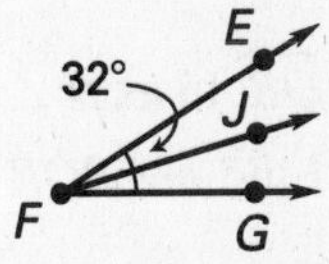

$\angle ABC$ is bisected by $\overrightarrow{BD}$. Find the value of x.

7.

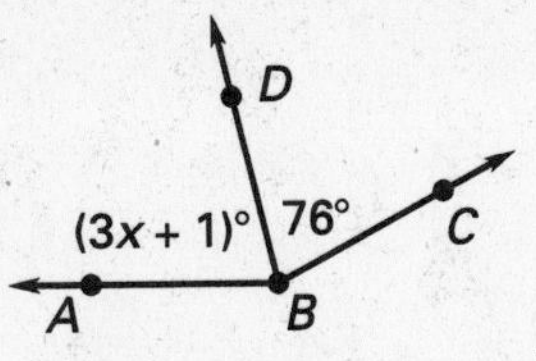

8.

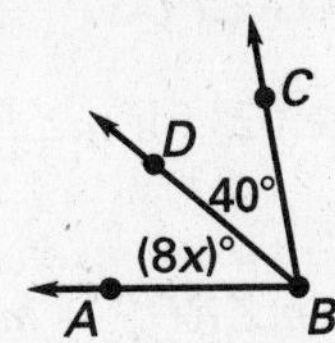

9.

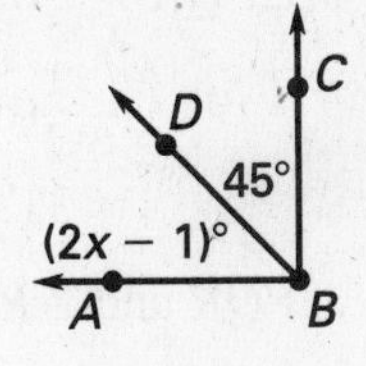

Use the figure at the right to complete the statement.

10. If $\angle KJM \cong \angle MJL$, then __?__ is an angle bisector.

11. If $\overrightarrow{JM}$ bisects $\angle KJL$, then $m\angle$ __?__ $= m\angle$ __?__.

12. If $\overrightarrow{JM}$ bisects $\angle KJL$ and $m\angle MJL = 48°$, then $m\angle$ __?__ $= 48°$.

13. If $\overrightarrow{JM}$ bisects $\angle KJL$ and $m\angle KJM = 46°$, then $m\angle KJL =$ __?__°.

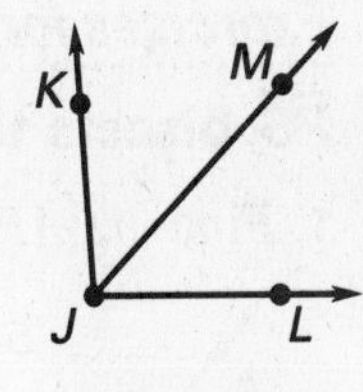

14. When an air hockey puck is hit into the sideboards, it bounces off so that $\angle 1$ and $\angle 2$ are congruent. Find $m\angle 1$ and $m\angle 2$.

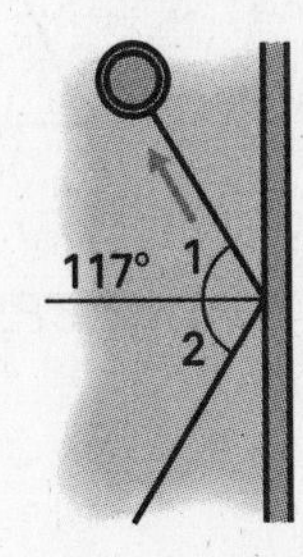

Lesson 2.2

Copyright © McDougal Littell Inc.
All rights reserved.

NAME ______________________________ DATE ____________

Reteaching with Practice

For use with pages 60–66

GOAL **Bisect an angle.**

VOCABULARY

An **angle bisector** is a ray that divides an angle into two angles that are congruent.

EXAMPLE 1

Find Angle Measures

$\overrightarrow{QR}$ bisects $\angle PQS$, and $m\angle PQS = 62°$.
Find $m\angle PQR$ and $m\angle RQS$.

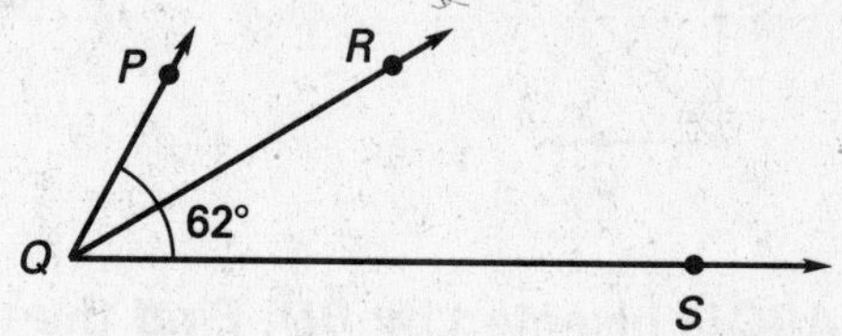

SOLUTION

$m\angle PQR = \frac{1}{2}(m\angle PQS)$ — $\overrightarrow{QR}$ bisects $\angle PQS$.

$= \frac{1}{2}(62°)$ — Substitute 62° for $m\angle PQS$.

$= 31°$ — Simplify.

$\angle PQR$ and $\angle RQS$ are congruent, so $m\angle RQS = m\angle PQR = 31°$.

So, $m\angle PQR = 31°$ and $m\angle RQS = 31°$.

Exercises for Example 1

$\overrightarrow{PS}$ bisects the angle. Find the angle measures.

1. Find $m\angle APS$ and $m\angle SPB$.

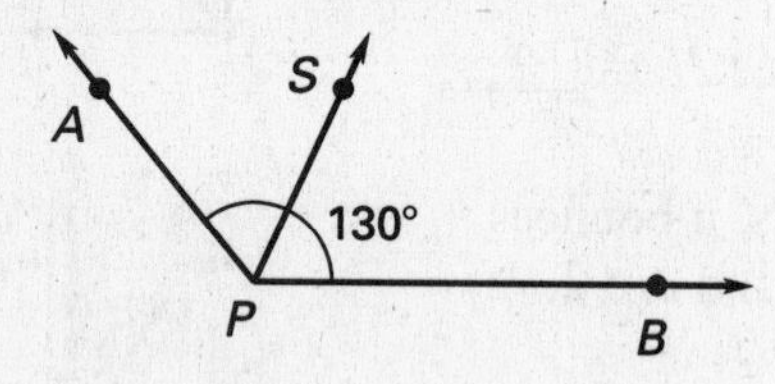

2. Find $m\angle CPS$ and $m\angle DPS$.

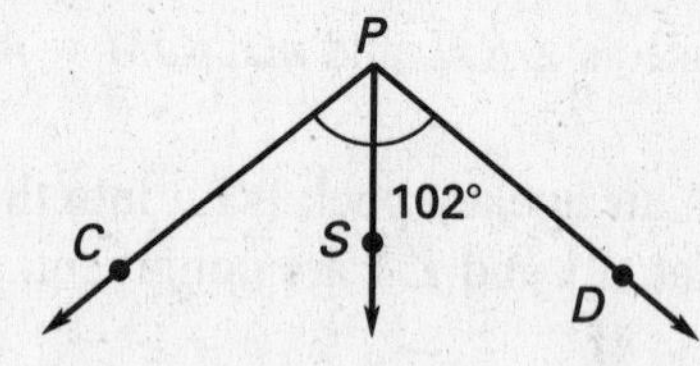

3. Find $m\angle EPS$ and $m\angle SPF$.

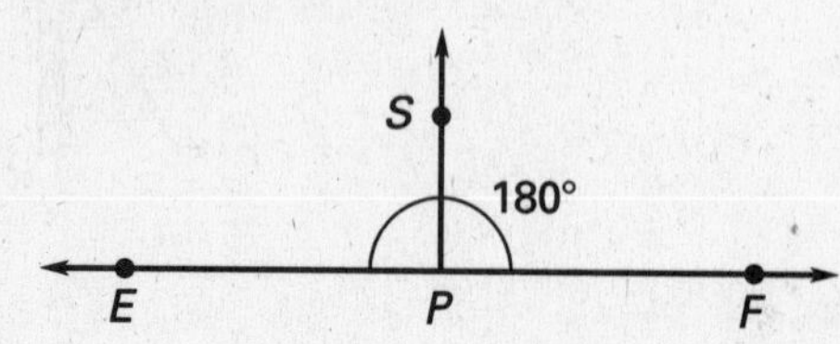

4. Find $m\angle GPS$ and $m\angle HPS$.

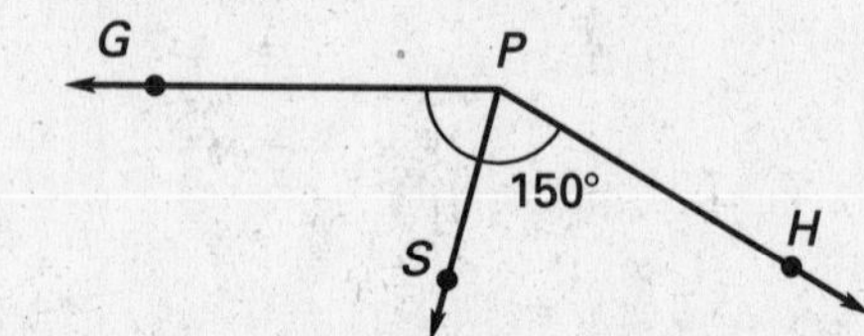

Copyright © McDougal Littell Inc.
All rights reserved.

LESSON 2.2 CONTINUED

NAME ______________________ DATE __________

Reteaching with Practice

For use with pages 60–66

EXAMPLE 2 Find Angle Measures and Classify an Angle

$\overrightarrow{BD}$ bisects $\angle ABC$, and $m\angle ABD = 45°$.

a. Find $m\angle CBD$ and $m\angle ABC$.

b. Determine whether $\angle ABC$ is *acute*, *right*, *obtuse*, or *straight*. Explain.

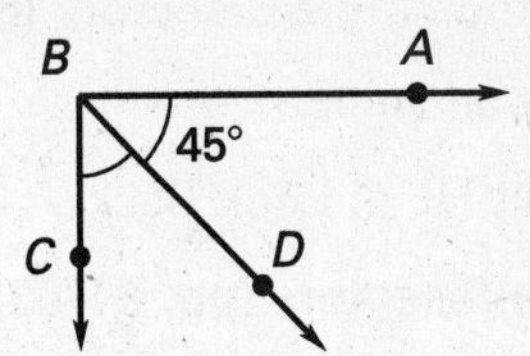

SOLUTION

a. $\overrightarrow{BD}$ bisects $\angle ABC$, so $m\angle CBD = m\angle ABD$.
You know that $m\angle ABD = 45°$.
Therefore, $m\angle CBD = 45°$.
The measure of $\angle ABC$ is twice the measure of $\angle ABD$.
$m\angle ABC = 2 \cdot (m\angle ABD) = 2 \cdot (45°) = 90°$
So, $m\angle CBD = 45°$ and $m\angle ABC = 90°$.

b. $\angle ABC$ is a right angle because its measure is 90°.

Exercises for Example 2

$\overrightarrow{BD}$ bisects $\angle ABC$. Find $m\angle CBD$ and $m\angle ABC$. Then determine whether $\angle ABC$ is *acute, right, obtuse,* or *straight*.

5.

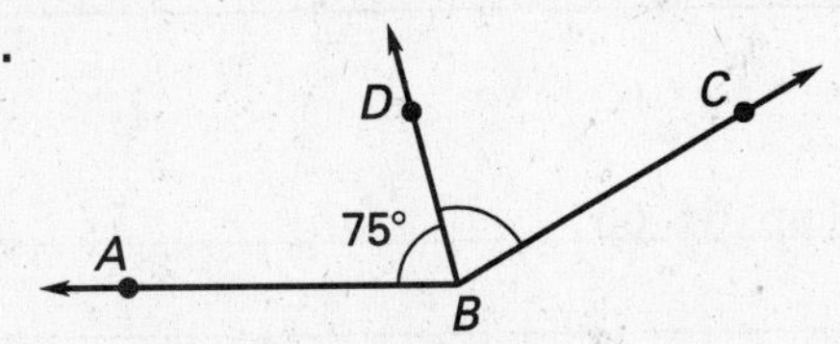

6.

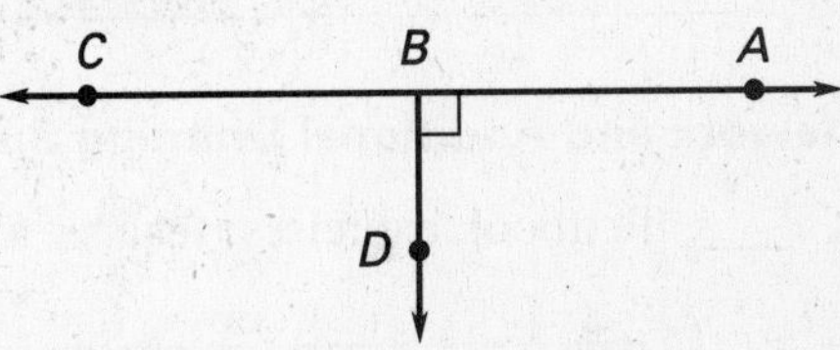

EXAMPLE 3 Use Algebra with Angle Measures

$\overrightarrow{BD}$ bisects $\angle ABC$. Find the value of x.

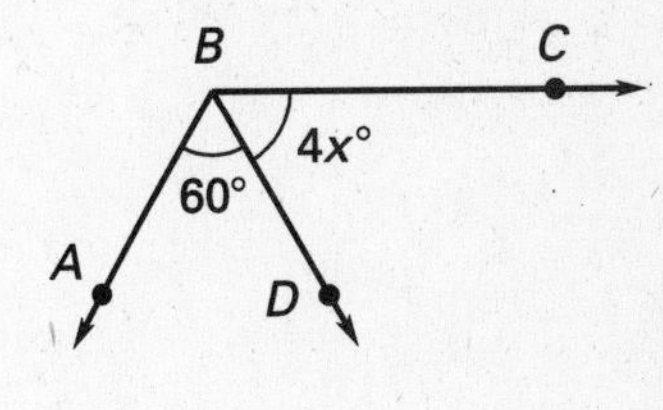

SOLUTION

$m\angle ABD = m\angle CBD$	$\overrightarrow{BD}$ bisects $\angle ABC$.
$60° = 4x°$	Substitute given measures.
$15 = x$	Divide each side by 4.

Exercises for Example 3

$\overrightarrow{BD}$ bisects $\angle ABC$. Find the value of x.

7. A D 45° $(x - 10)°$ B C

8.

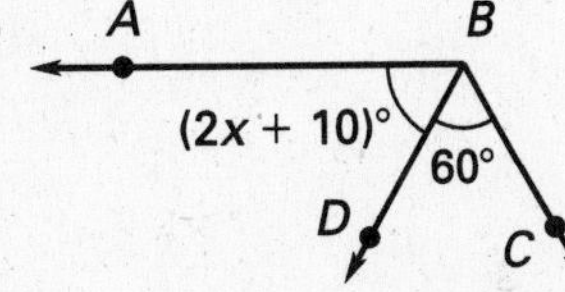

9.

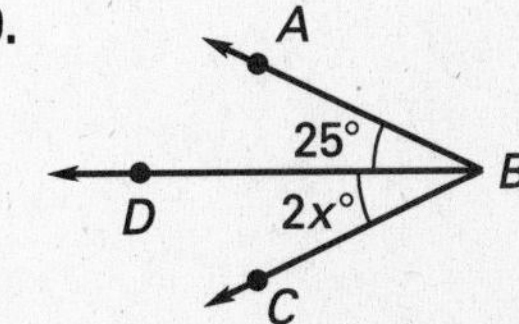

Copyright © McDougal Littell Inc.
All rights reserved.

LESSON 2.2

NAME ______________________________ DATE __________

Quick Catch-Up for Absent Students

For use with pages 60–66

The items checked below were covered in class on (date missed) __________

Activity 2.2: Folding Angle Bisectors (p. 60)

____ **Goal:** Bisect an angle.

____ Student Help: Look Back

Lesson 2.2: Angle Bisectors (pp. 61–63)

____ **Goal:** Bisect an angle.

Material Covered:

____ Example 1: Find Angle Measures

____ Example 2: Find Angle Measures and Classify an Angle

____ Example 3: Use Angle Bisectors

____ Example 4: Use Algebra with Angle Measures

Vocabulary:

angle bisector, p. 61

____ Other (specify) ______________________________

____

Homework and Additional Learning Support

____ Textbook exercises (teacher to specify) pp. 64–66

____

____ Internet: More Examples at classzone.com

____ *Reteaching with Practice* worksheet

Copyright © McDougal Littell Inc.
All rights reserved.

LESSON 2.2

NAME ______________________________ DATE __________

Real-Life Application: When Will I Ever Use This?

For use with pages 60–66

Billiards

When playing billiards, one of the most difficult skills to master is the *bank shot.* In this shot, you do not shoot directly at the pocket. Instead, you ricochet the ball off of a cushion, or side of the table, so that it goes into a pocket. To master a bank shot, you must know that the angle of the ball hitting the cushion is equal to the angle of the ball leaving the cushion. In the figure at the right, $\angle 1 \cong \angle 2$, so line b bisects the angle formed by the shot.

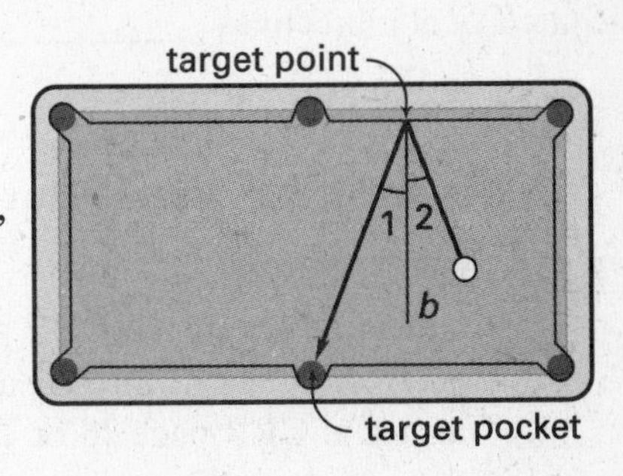

To find the angle bisector of the shot, and therefore, the point at which to hit the bank shot, follow the steps below.

1. In your mind, draw a line (line p) from the ball to the cushion of the side opposite the target pocket so that it forms a right angle. Then draw a line (line t) from that point to the target pocket. Then draw a line (line s) from the ball to the side pocket opposite the target pocket.

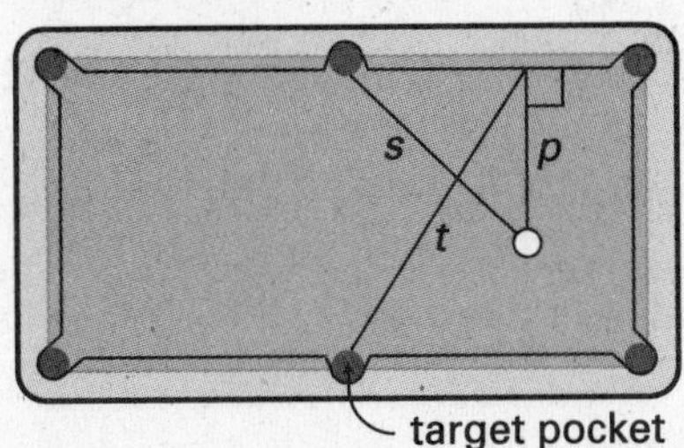

2. Draw a line (line b) that goes through the intersection of lines s and t so that it meets the table at a right angle. The point at which this line hits the table is the target point. Line b is an angle bisector of the path of the ball, which is shown by the dashed line segments.

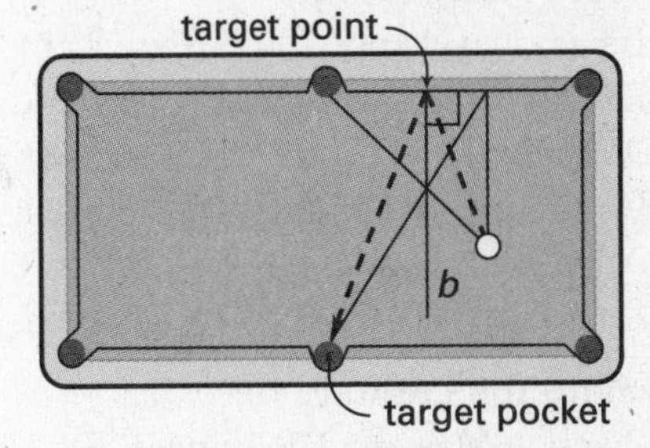

Given the position of the ball and the target pocket, show how to bank the ball using the steps above. Then draw the path of the ball using a ruler. Measure the angle formed by the path of the ball. Does line *b* bisect this angle?

1.

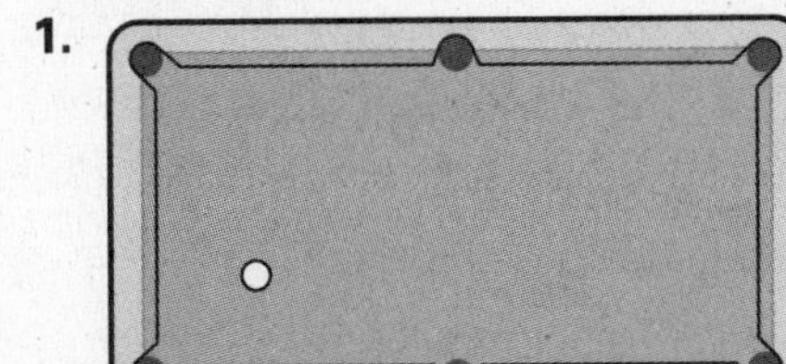

2.

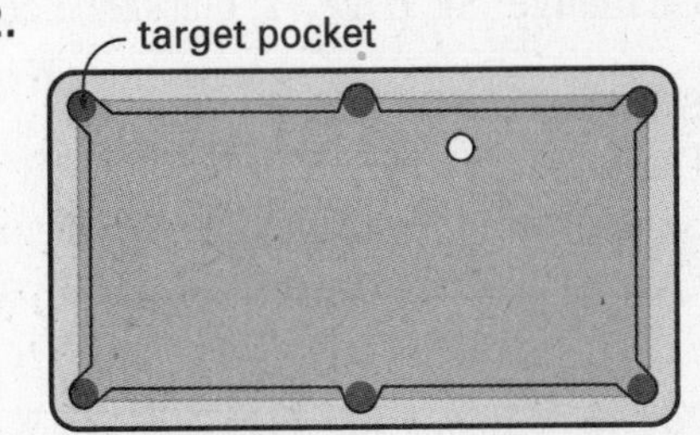

Lesson 2.2

Copyright © McDougal Littell Inc.
All rights reserved.

LESSON 2.3

TEACHER'S NAME ____________ CLASS ______ ROOM ______ DATE ______

Lesson Plan

2-day lesson (See *Pacing the Chapter,* TE page 50A) For use with pages 67–73

GOAL Find measures of complementary and supplementary angles.

State/Local Objectives ____________

✓ **Check the items you wish to use for this lesson.**

STARTING OPTIONS

____ Homework Check (2.2): TE page 64; Answer Transparencies
____ Homework Quiz (2.2): TE page 66, CRB page 28, or Transparencies
____ Warm-Up: CRB page 28 or Transparencies

TEACHING OPTIONS

____ Examples: Day 1: 1–2, SE pages 67–68; Day 2: 3–4, SE pages 68–69
____ Extra Examples: TE pages 68–69
____ Checkpoint Exercises: Day 1: 1–3, SE page 67; Day 2: 4–6, SE pages 68–69
____ Technology Keystrokes for Ex. 39 on SE page 72: CRB page 29
____ Concept Check: TE page 69
____ Guided Practice Exercises: Day 1: 1–5, SE page 70; Day 2: 6–7, SE page 70
____ Visualize It! Transparencies: 7–9

APPLY/HOMEWORK

Homework Assignment

____ Basic: Day 1: pp. 70–73 Exs. 8–14, 29–32, 44–59
Day 2: SRH p. 673 Exs. 13–17 odd; pp. 70–73 Exs. 15–27 odd, 28, 33, 40–42, Quiz 1
____ Average: Day 1: pp. 70–73 Exs. 8–14, 29–32, 44, 46–59
Day 2: pp. 70–73 Exs. 16–28 even, 34–42 even, 45, Quiz 1
____ Advanced: Day 1: pp. 70–73 Exs. 8–14 even, 29–32, 45–59
Day 2: pp. 70–73 Exs. 16–28 even, 33–41 odd, 43*, Quiz 1; EC: classzone.com

Reteaching the Lesson

____ Practice Masters: CRB pages 30–31 (Level A, Level B)
____ Reteaching with Practice: CRB pages 32–33 or Practice Workbook with Examples; Resources in Spanish

Extending the Lesson

____ Real-Life Application: CRB page 35
____ Challenge: SE page 72; classzone.com

ASSESSMENT OPTIONS

____ Daily Quiz (2.3): TE page 73, CRB page 39, or Transparencies
____ Standardized Test Practice: SE page 73; Transparencies
____ Quiz (2.1–2.3): SE page 73; CRB page 36

Notes ____________

Copyright © McDougal Littell Inc.
All rights reserved.

TEACHER'S NAME ____________________ CLASS ______ ROOM ______ DATE ______

Lesson Plan for Block Scheduling

1-block lesson (See *Pacing the Chapter,* TE page 50A)

For use with pages 67–73

GOAL **Find measures of complementary and supplementary angles.**

State/Local Objectives ____________________

____

____

CHAPTER PACING GUIDE	
Day	**Lesson**
1	2.1
2	2.2
3	**2.3**
4	2.4
5	2.5
6	2.6
7	Ch. 2 Review and Assess

✓ Check the items you wish to use for this lesson.

STARTING OPTIONS

____ Homework Check (2.2): TE page 64; Answer Transparencies
____ Homework Quiz (2.2): TE page 66, CRB page 28, or Transparencies
____ Warm-Up: CRB page 28 or Transparencies

TEACHING OPTIONS

____ Examples: 1–4, SE pages 67–69
____ Extra Examples: TE pages 68–69
____ Checkpoint Exercises: 1–6, SE pages 67–69
____ Technology Keystrokes for Ex. 39 on SE page 72: CRB page 29
____ Concept Check: TE page 69
____ Guided Practice Exercises: 1–7, SE page 70
____ Visualize It! Transparencies: 7–9

APPLY/HOMEWORK

Homework Assignment

____ Block Schedule: pp. 70–73 Exs. 8–14, 16–28 even, 29–32, 34–42 even, 44–59, Quiz 1

Reteaching the Lesson

____ Practice Masters: CRB pages 30–31 (Level A, Level B)
____ Reteaching with Practice: CRB pages 32–33 or Practice Workbook with Examples; Resources in Spanish

Extending the Lesson

____ Real-Life Application: CRB page 35
____ Challenge: SE page 72; classzone.com

ASSESSMENT OPTIONS

____ Daily Quiz (2.3): TE page 73, CRB page 39, or Transparencies
____ Standardized Test Practice: SE page 73; Transparencies
____ Quiz (2.1–2.3): SE page 73; CRB page 36

Notes ____________________

____

____

Copyright © McDougal Littell Inc.
All rights reserved.

LESSON 2.3

NAME ______________________________ DATE ____________

Available as a transparency

WARM-UP EXERCISES

For use before Lesson 2.3, pages 67–73

Name an example of each type of angle from the figure at the right.

1. obtuse
2. right
3. acute
4. straight

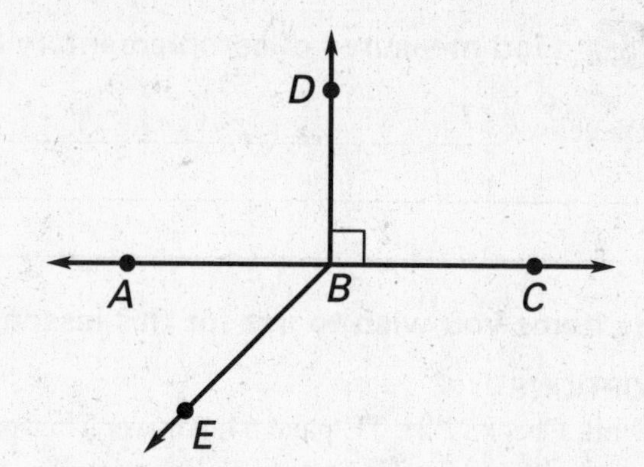

DAILY HOMEWORK QUIZ

For use after Lesson 2.2, pages 60–66

1. In the figure at the right, $\overrightarrow{BD}$ bisects $\angle ABC$. Find $m\angle ABD$ and $m\angle DBC$.

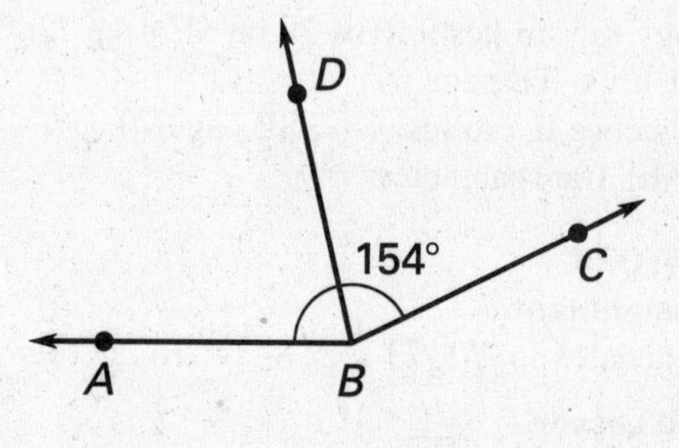

2. In the figure at the right, $\overrightarrow{HL}$ bisects $\angle GHJ$. Find $m\angle GHL$ and $m\angle GHJ$. Then determine whether $\angle GHJ$ is *acute*, *right*, *obtuse*, or *straight*.

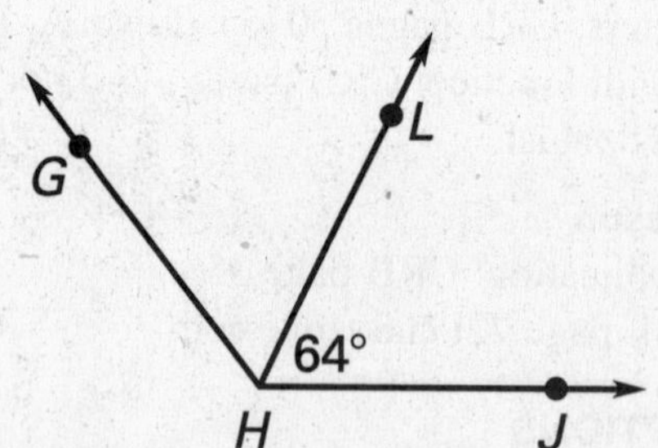

$\overrightarrow{QS}$ bisects $\angle PQR$. Find the value of the variable.

3.

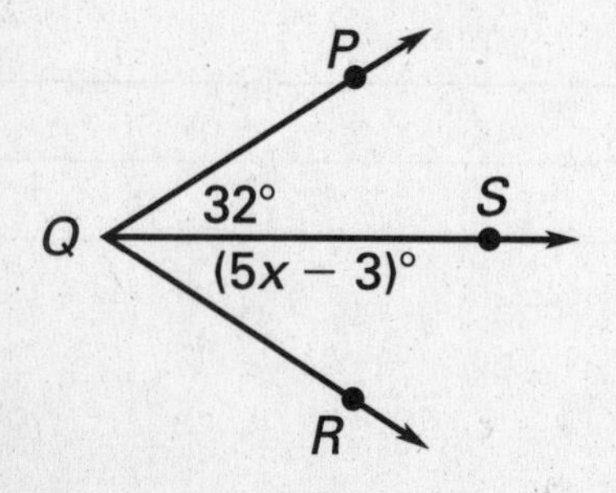

4.

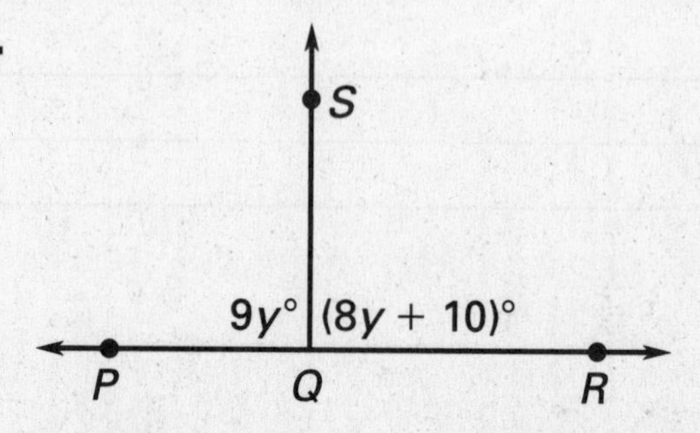

Lesson 2.3

Copyright © McDougal Littell Inc.
All rights reserved.

NAME ______________________ DATE ________

Technology Keystrokes

For use with Exercise 39, page 72

Keystrokes for Exercise 39

TI-92

1. Draw $\overleftrightarrow{AD}$.

   F2 4 (Move cursor to desired location.) ENTER *A*

   (Move cursor to location for point *D*.) ENTER

   Draw point *D*. F2 2 ENTER *D*

2. Draw $\overleftrightarrow{BC}$.

   F2 4 (Move cursor to desired location.) ENTER *B*

   (Move cursor to location for point *C*.) ENTER

   Draw point *C*. F2 2 ENTER *C*

3. Draw the intersection point of $\overleftrightarrow{AD}$ and $\overleftrightarrow{BC}$.

   F2 3 (Move cursor to $\overleftrightarrow{AD}$.) ENTER (Move cursor to $\overleftrightarrow{BC}$.) ENTER *Q*

4. Measure three angles.

   F6 3 (Move cursor to desired side.) ENTER

   (Move cursor to point *Q*.) ENTER

   (Move cursor to desired side.) ENTER

   Repeat this step to find the measures of two other angles.

5. Drag the points.

   (Move cursor to desired point.) Use the drag key [drag key] and the cursor pad to drag the point.

SKETCHPAD

1. Draw $\overleftrightarrow{AD}$. Choose the line tool. Draw $\overleftrightarrow{AB}$. Choose the text tool. Double click on the label *B*, type *D*, and click OK.
2. Draw $\overleftrightarrow{BC}$. Choose the line tool. Draw $\overleftrightarrow{CD}$. Choose the text tool. Double click on the label *C*, type *B*, and click OK. Double click on the label *D*, type *C*, and click OK.
3. Draw the intersection point. Choose the selection arrow tool. Select $\overleftrightarrow{AD}$, hold down the shift key and select $\overleftrightarrow{BC}$. Choose **Point At Intersection** from the **Construct** menu. Choose the text tool. Double click on the intersection point label, type *Q*, click OK.
4. Measure three angles. Choose the selection arrow tool. Select three points that make up an angle (hold down the shift key when selecting the vertex and final point). Choose **Angle** from the **Measure** menu. Repeat this procedure for the other two angles.
5. Drag the points. Choose the selection arrow tool. Drag the points.

Copyright © McDougal Littell Inc.
All rights reserved.

LESSON 2.3

NAME ______________________________ DATE __________

Practice A

For use with pages 67–73

Decide whether the statement is *true* or *false*. If the statement is false, reword the statement so that the statement is true.

1. Two angles are complementary if the sum of their measures is 180°.
2. Two angles are supplementary if the sum of their measures is 180°.
3. Two angles are adjacent angles if they share a common vertex.
4. A theorem is a true statement that follows from other true statements.

Determine whether the angles are *complementary, supplementary,* or *neither*.

5. 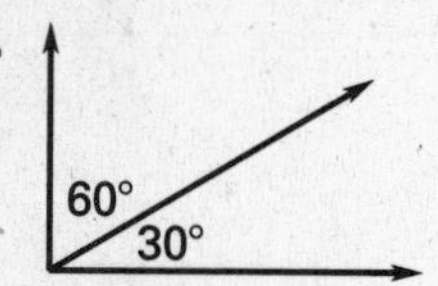

6. 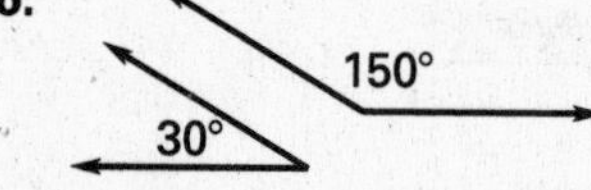

7. 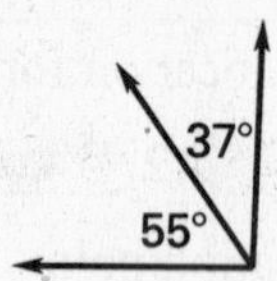

Find the measure of a complement of the angle given.

8. 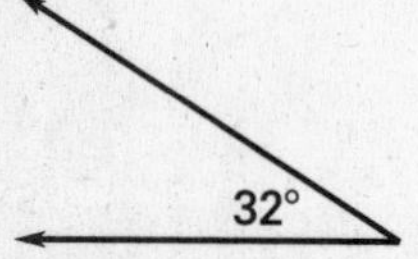

9. 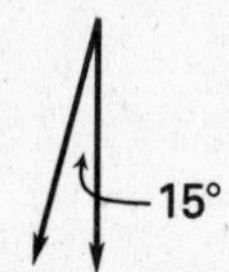

10. 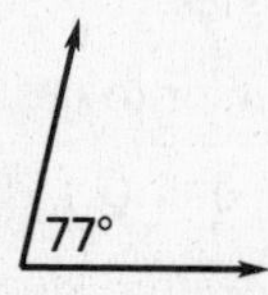

Find the measure of a supplement of the angle given.

11. 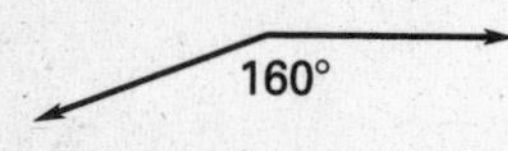

12. 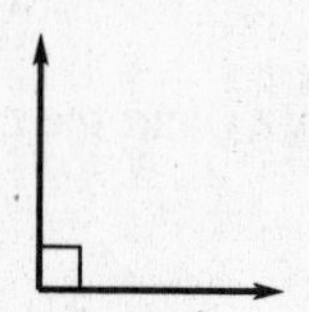

13. 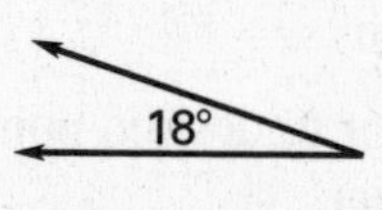

Use the diagram to complete the statement.

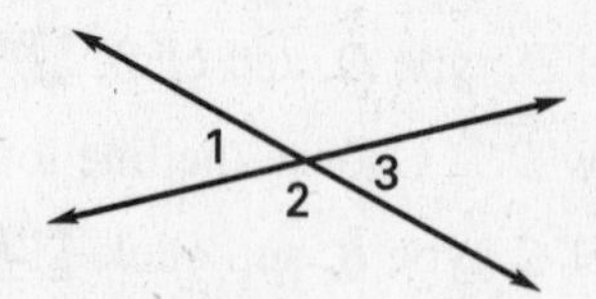

14. ∠1 and __?__ are supplementary angles.
15. ∠3 and __?__ are supplementary angles.
16. __?__ ≅ __?__ by the Congruent Supplements Theorem.

17. The foul lines of a baseball field intersect at home plate to form a right angle, $\angle THF$. You hit a baseball whose path forms an angle of 30° with the third base foul line. What is the measure of the angle formed by the first base foul line and the path of the ball?

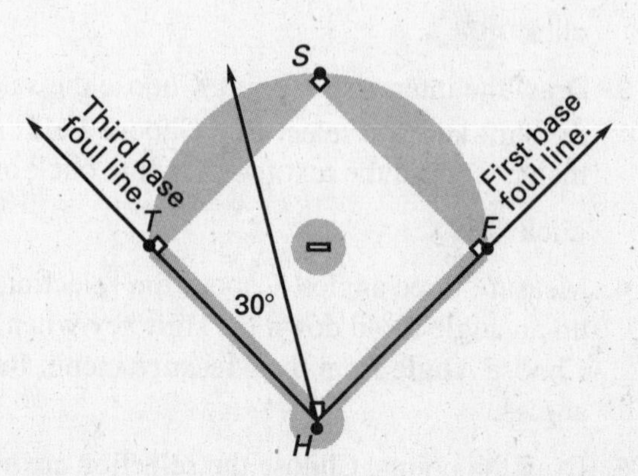

Copyright © McDougal Littell Inc.
All rights reserved.

LESSON 2.3

NAME ______________________________ DATE __________

Practice B

For use with pages 67–73

Complete the statement.

1. Two angles are complementary if the sum of their measures is __?__°.
2. Two angles are supplementary if the sum of their measures is __?__°.
3. If two angles share a common vertex and side, but have no common interior points, then the two angles are __?__ angles.
4. A true statement that follows from other true statements is called a __?__.

Determine whether the angles whose measures are given are *complementary, supplementary,* or *neither.* Also tell whether the angles are *adjacent* or *nonadjacent.*

5.
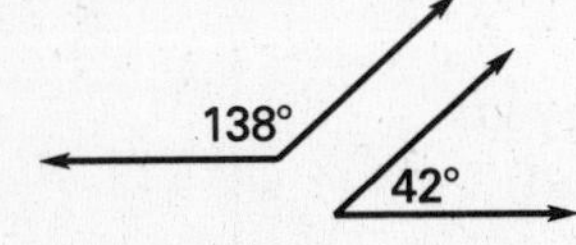

6.
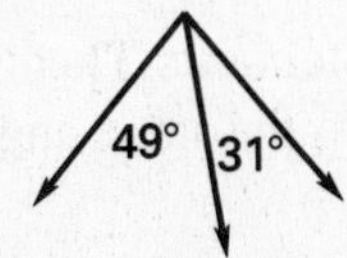

7.
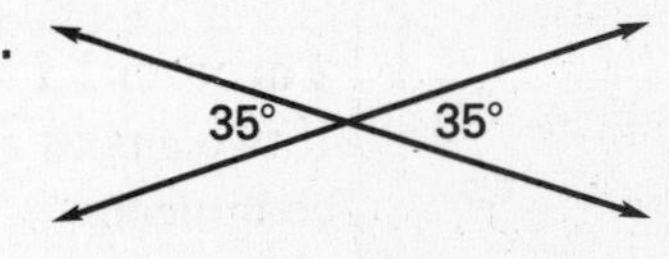

Find the measure of a complement of the angle.

8. $m\angle Y = 40°$
9. $m\angle K = 12°$
10. $m\angle P = 64°$
11. $m\angle T = 85°$

Find the measure of a supplement of the angle.

12. $m\angle A = 54°$
13. $m\angle R = 115°$
14. $m\angle Z = 22°$
15. $m\angle F = 90°$

$\angle PQS$ and $\angle SQR$ are complementary angles. Find the value of the variable.

16.
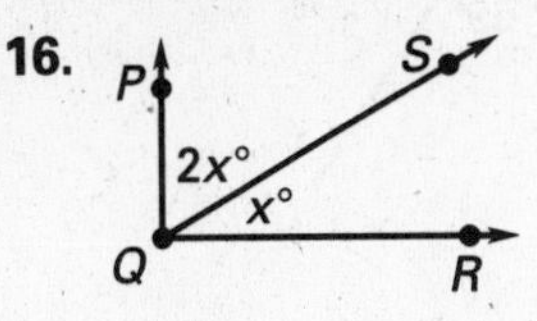

17.
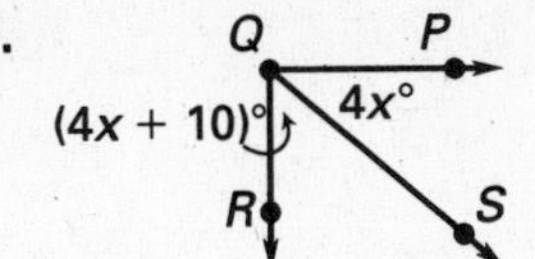

18.
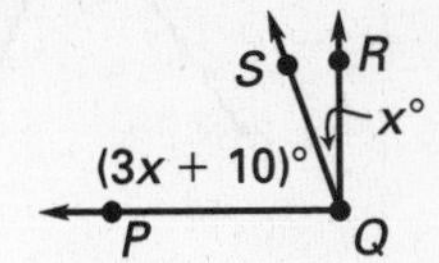

19. $\angle ABC$ and $\angle DBE$ are right angles. Name an angle that is congruent to $\angle 3$. Explain.

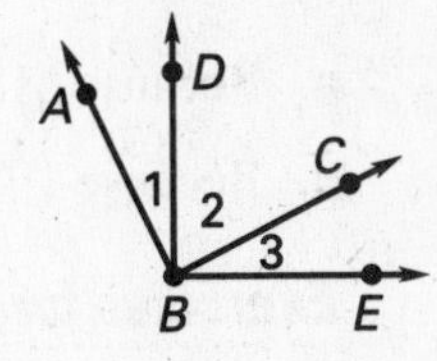

In Exercises 20 and 21, use the drawing of a teeter-totter.

20. The marked angles are supplementary. Find the value of x.
21. By how many degrees would the angle of the teeter-totter have to change so that it forms a right angle with its vertical support bar? (Hint: Find the measure of a complement of a 74° angle.)

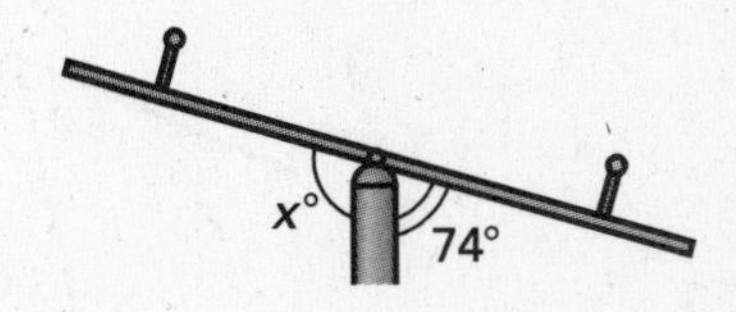

Lesson 2.3

Copyright © McDougal Littell Inc.
All rights reserved.

LESSON 2.3

NAME ______________________ DATE ________

Reteaching with Practice

For use with pages 67–73

GOAL **Find measures of complementary and supplementary angles.**

VOCABULARY

Two angles are **complementary angles** if the sum of their measures is 90°. Each angle is the **complement** of the other.

Two angles are **supplementary angles** if the sum of their measures is 180°. Each angle is the **supplement** of the other.

Two angles are **adjacent angles** if they share a common vertex and side, but have no common interior points.

A **theorem** is a true statement that follows from other true statements.

Theorem 2.1 Congruent Complements Theorem
If two angles are complementary to the same angle, then they are congruent.

Theorem 2.2 Congruent Supplements Theorem
If two angles are supplementary to the same angle, then they are congruent.

EXAMPLE 1 *Identify Complements and Supplements*

Determine whether the angles are *complementary*, *supplementary*, or *neither*.

a.

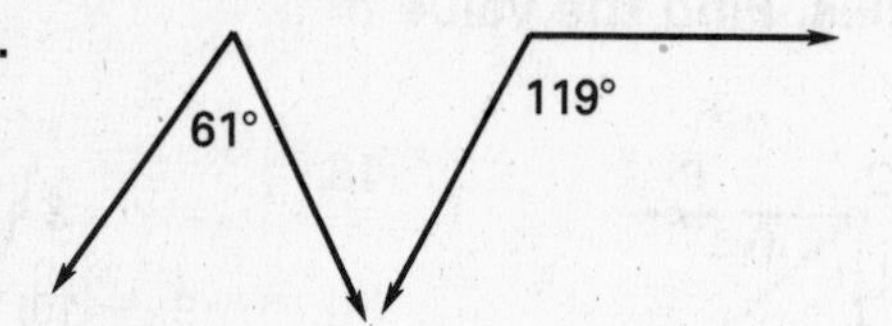

b.

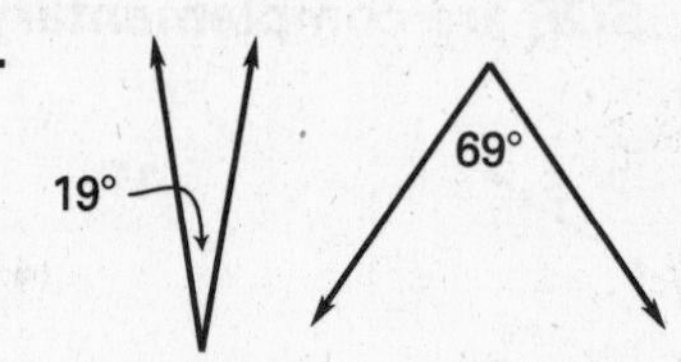

SOLUTION

a. Because 61° + 119° = 180°, the angles are supplementary.

b. Because 19° + 69° = 88°, the angles are neither complementary nor supplementary.

Exercises for Example 1

Determine whether the angles are *complementary, supplementary,* or *neither.*

1.

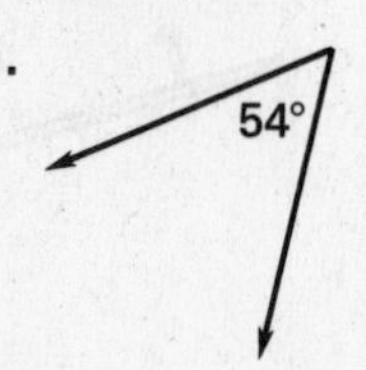

2.

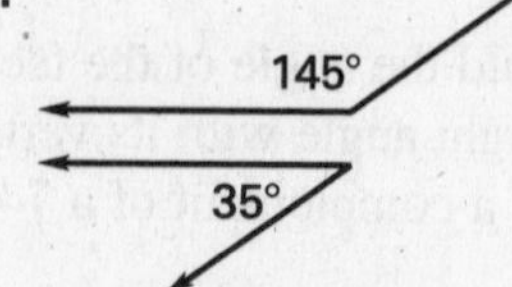

3.

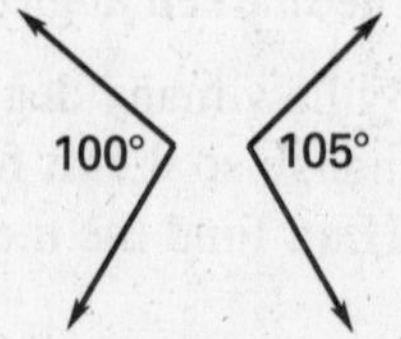

Lesson 2.3

Copyright © McDougal Littell Inc.
All rights reserved.

NAME ______________________________ DATE __________

Reteaching with Practice

For use with pages 67–73

EXAMPLE 2 Identify Adjacent Angles

Tell whether the numbered angles are *adjacent* or *nonadjacent*.

a.

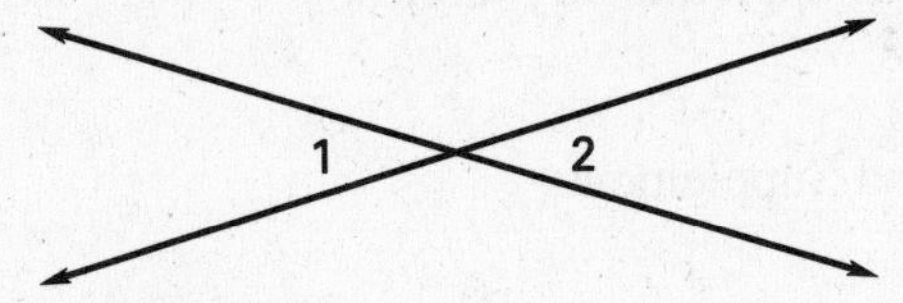

b.

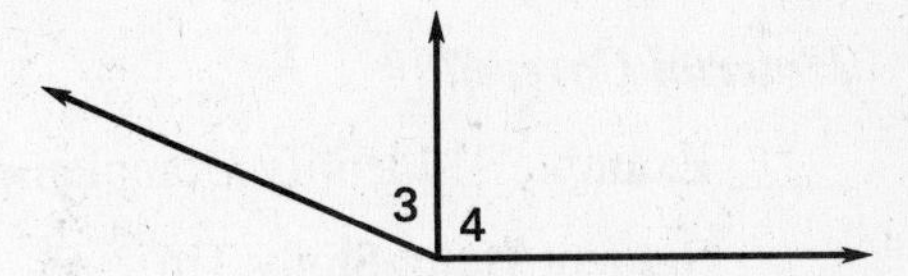

SOLUTION

a. Although $\angle 1$ and $\angle 2$ share a common vertex, they do not share a common side. Therefore, $\angle 1$ and $\angle 2$ are nonadjacent.

b. Because $\angle 3$ and $\angle 4$ share a common vertex and side, and do not have any common interior points, $\angle 3$ and $\angle 4$ are adjacent.

Exercises for Example 2

Tell whether the numbered angles are *adjacent* or *nonadjacent*.

4.

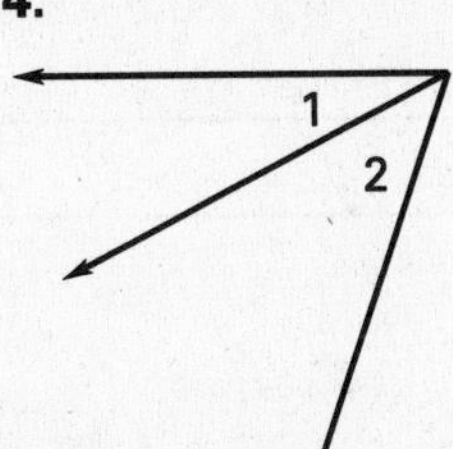

5.

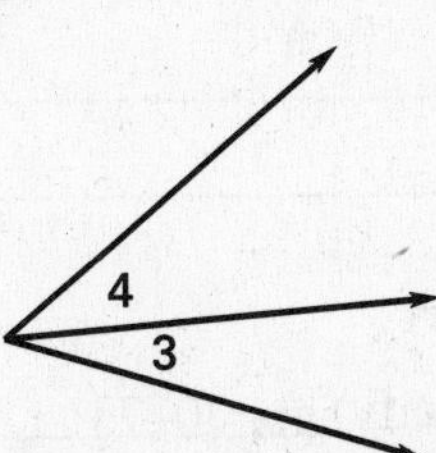

6.

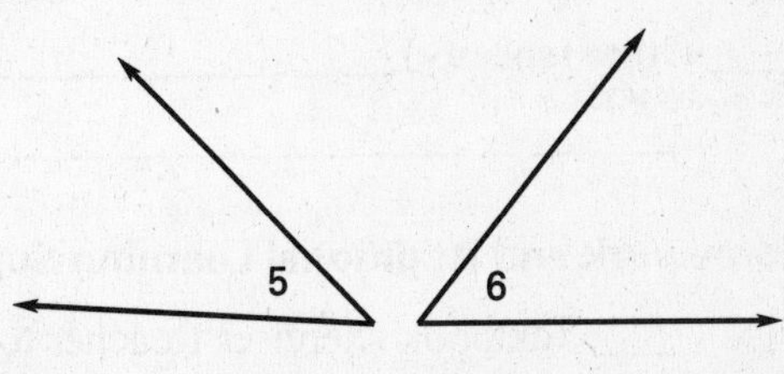

EXAMPLE 3 Measures of Complements and Supplements

a. $\angle A$ is a supplement of $\angle B$, and $m\angle B = 42°$. Find $m\angle A$.

b. $\angle C$ is a complement of $\angle D$, and $m\angle C = 42°$. Find $m\angle D$.

SOLUTION

a. $\angle A$ and $\angle B$ are supplements, so their sum is 180°.

$$m\angle A + m\angle B = 180°$$
$$m\angle A + 42° = 180°$$
$$m\angle A + 42° - 42° = 180° - 42°$$
$$m\angle A = 138°$$

b. $\angle C$ and $\angle D$ are complements, so their sum is 90°.

$$m\angle C + m\angle D = 90°$$
$$42° + m\angle D = 90°$$
$$42° + m\angle D - 42° = 90° - 42°$$
$$m\angle D = 48°$$

Exercises for Example 3

Find the angle measure.

7. $\angle A$ is a complement of $\angle B$, and $m\angle A = 11°$. Find $m\angle B$.

8. $\angle A$ is a supplement of $\angle B$, and $m\angle B = 122°$. Find $m\angle A$.

9. $\angle C$ is a complement of $\angle D$, and $m\angle C = 88°$. Find $m\angle D$.

Copyright © McDougal Littell Inc.
All rights reserved.

LESSON 2.3

NAME ______________________________ DATE __________

Quick Catch-Up for Absent Students

For use with pages 67–73

The items checked below were covered in class on (date missed) __________

Lesson 2.3: Complementary and Supplementary Angles (pp. 67–69)

____ **Goal:** Find measures of complementary and supplementary angles.

Material Covered:

____ Example 1: Identify Complements and Supplements

____ Student Help: Study Tip

____ Example 2: Identify Adjacent Angles

____ Example 3: Measures of Complements and Supplements

____ Student Help: Visual Strategy

____ Example 4: Use a Theorem

Vocabulary:

complementary angles, p. 67
complement of an angle, p. 67
supplementary angles, p. 67
supplement of an angle, p. 67
adjacent angles, p. 68
theorem, p. 69

____ Other (specify) ______________________________

____

Homework and Additional Learning Support

____ Textbook exercises (teacher to specify) pp. 70–73

____

____ Internet: Homework Help at classzone.com

____ *Reteaching with Practice* worksheet

Lesson 2.3

Copyright © McDougal Littell Inc.
All rights reserved.

NAME ______________________ DATE __________

Real-Life Application: When Will I Ever Use This?

For use with pages 67–73

Wallpaper Border

When applying wallpaper border around a window or a door to create a distinctive decorative frame, you need to use congruent angles.

You are wallpapering a border around a rectangular window frame. As shown in the diagram at the right, you need to make perfect miter joints at the corners.

The best method to make the corners match would be to make sure that the most prominent design on the border would not be at the corner. Then let the two strips overlap, matching the designs so they look professional. Now take a straightedge and cut through both layers with a knife. Peel away the excess. Do the same on the other corner, making sure your cuts match.

In Exercises 1–5, use the diagram.

1. What is $m\angle 1 + m\angle 3$? Explain your reasoning.
2. Do $\angle 1$ and $\angle 3$ have to be equal in measure? $\angle 2$ and $\angle 4$? Explain your reasoning.
3. If $m\angle 3 = 50°$, what is $m\angle 1$?
4. If $m\angle 3 = 52°$, and $m\angle 1 = m\angle 2$, what is $m\angle 4$, $m\angle 1$, and $m\angle 2$?
5. What are the benefits of using the method described for applying a wallpaper border?

Copyright © McDougal Littell Inc.
All rights reserved.

LESSON 2.3

NAME ______________________________ DATE __________

Quiz 1

For use after Lessons 2.1–2.3

E is the midpoint of $\overline{DF}$. Find the segment lengths.

1. Find DE and EF.

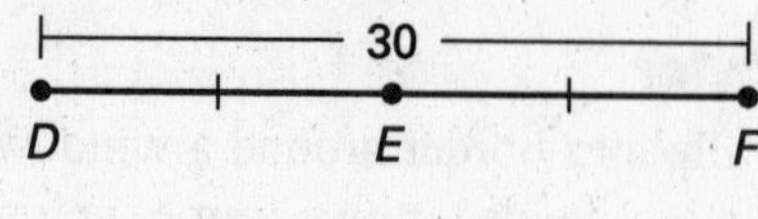

2. Find DE and DF.

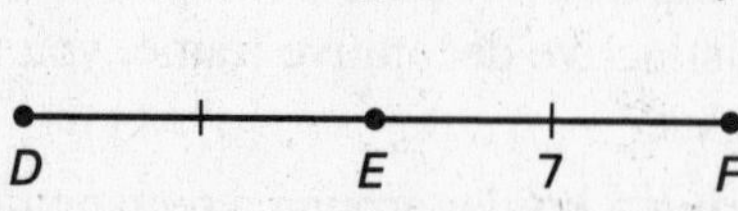

3. B is the midpoint of $\overline{AC}$. Find the value of x.

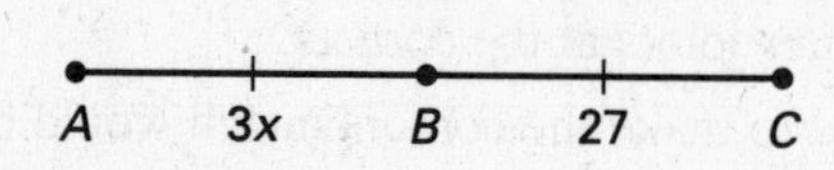

Find the coordinates of the midpoint of $\overline{AB}$.

4. $A(7, -4)$, $B(5, 2)$

5. $A(-3, -2)$, $B(9, 6)$

$\overrightarrow{BD}$ bisects $\angle ABC$. Find the angle measure or variable.

6. Find $m\angle ABD$.

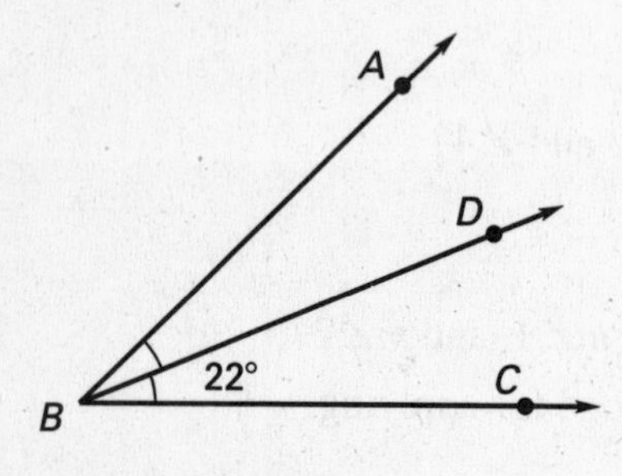

7. Find $m\angle DBC$.

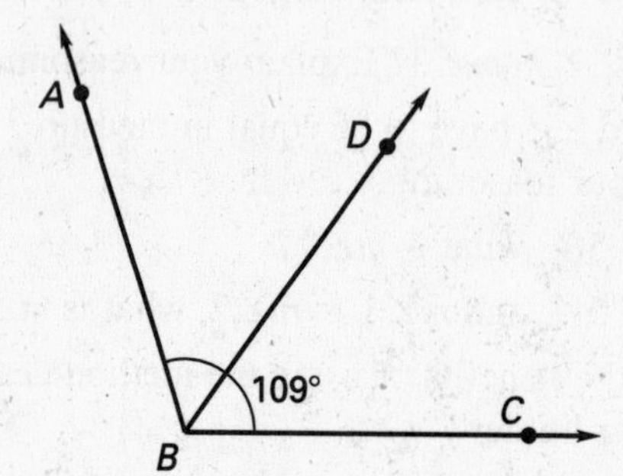

8. Find the value of x.

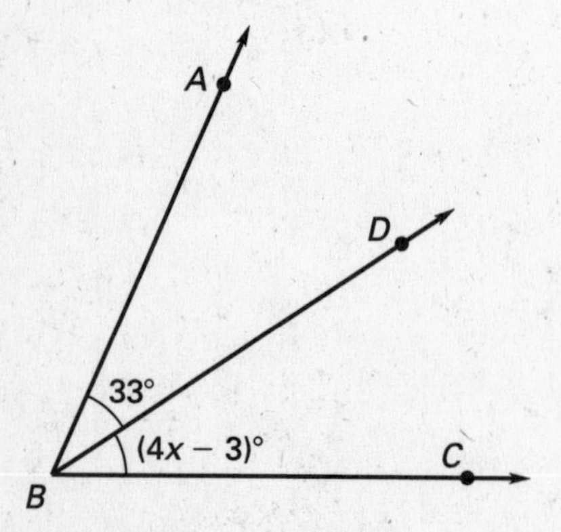

In Exercises 9–11, determine whether $\angle F$ and $\angle G$ are *complementary, supplementary,* or *neither.*

9. $m\angle F = 29°$, $m\angle G = 61°$

10. $m\angle F = 159°$, $m\angle G = 11°$

11. $m\angle F = 37°$, $m\angle G = 37°$

12. The measure of $\angle M$ is 56°. Find the measure of a complement and a supplement of a $\angle M$.

Answers

1. ______________
2. ______________
3. ______________
4. ______________
5. ______________
6. ______________
7. ______________
8. ______________
9. ______________
10. ______________
11. ______________
12. ______________

Lesson 2.3

Copyright © McDougal Littell Inc.
All rights reserved.

LESSON 2.4

TEACHER'S NAME ______________________ CLASS ______ ROOM ______ DATE ______

Lesson Plan

2-day lesson (See *Pacing the Chapter,* TE page 50A)

For use with pages 74–81

GOAL Find the measures of angles formed by intersecting lines.

State/Local Objectives ____

____

✓ **Check the items you wish to use for this lesson.**

STARTING OPTIONS

____ Homework Check (2.3): TE page 70; Answer Transparencies
____ Homework Quiz (2.3): TE page 73, CRB page 39, or Transparencies
____ Warm-Up: CRB page 39 or Transparencies

TEACHING OPTIONS

____ Activity: SE page 74
____ Examples: Day 1: 1–2, SE pages 75–76; Day 2: 3–5, SE pages 76–77
____ Extra Examples: TE pages 76–77
____ Checkpoint Exercises: Day 2: 1–6, SE page 77
____ Concept Check: TE page 77
____ Guided Practice Exercises: Day 1: 1–3, SE page 78; Day 2: 4–8, SE page 78
____ Visualize It! Transparencies: 10

APPLY/HOMEWORK

Homework Assignment

____ Basic: Day 1: pp. 78–81 Exs. 9–19, 24–27, 60–70 even
Day 2: pp. 78–81 Exs. 20–22, 28–35, 51–53, 59, 65–71 odd
____ Average: Day 1: pp. 78–81 Exs. 9–19, 23–27, 58, 64–71
Day 2: pp. 78–81 Exs. 20–22, 28–40, 51–54, 59–63
____ Advanced: Day 1: pp. 78–81 Exs. 9–19 odd, 23–27, 58–71
Day 2: pp. 78–81 Exs. 20–22, 28–40 even, 41–49 odd, 54–56, 57*; EC: classzone.com

Reteaching the Lesson

____ Practice Masters: CRB pages 40–41 (Level A, Level B)
____ Reteaching with Practice: CRB pages 42–43 or Practice Workbook with Examples; Resources in Spanish

Extending the Lesson

____ Challenge: SE page 81; classzone.com

ASSESSMENT OPTIONS

____ Daily Quiz (2.4): TE page 81, CRB page 47, or Transparencies
____ Standardized Test Practice: SE page 81; Transparencies

Notes ____

____

____

Copyright © McDougal Littell Inc.
All rights reserved.

LESSON 2.4

TEACHER'S NAME ______________ CLASS ______ ROOM ______ DATE ______

Lesson Plan for Block Scheduling

1-block lesson (See *Pacing the Chapter,* TE page 50A) For use with pages 74–81

GOAL Find the measures of angles formed by intersecting lines.

State/Local Objectives ______________________________

____

____

CHAPTER PACING GUIDE	
Day	**Lesson**
1	2.1
2	2.2
3	2.3
4	**2.4**
5	2.5
6	2.6
7	Ch. 2 Review and Assess

✓ Check the items you wish to use for this lesson.

STARTING OPTIONS

____ Homework Check (2.3): TE page 70; Answer Transparencies
____ Homework Quiz (2.3): TE page 73, CRB page 39, or Transparencies
____ Warm-Up: CRB page 39 or Transparencies

TEACHING OPTIONS

____ Activity: SE page 74
____ Examples: 1–5, SE pages 75–77
____ Extra Examples: TE pages 76–77
____ Checkpoint Exercises: 1–6, SE page 77
____ Concept Check: TE page 77
____ Guided Practice Exercises: 1–8, SE page 78
____ Visualize It! Transparencies: 10

APPLY/HOMEWORK

Homework Assignment

____ Block Schedule: pp. 78–81 Exs. 9–40, 51–54, 58–71

Reteaching the Lesson

____ Practice Masters: CRB pages 40–41 (Level A, Level B)
____ Reteaching with Practice: CRB pages 42–43 or Practice Workbook with Examples; Resources in Spanish

Extending the Lesson

____ Challenge: SE page 81; classzone.com

ASSESSMENT OPTIONS

____ Daily Quiz (2.4): TE page 81, CRB page 47, or Transparencies
____ Standardized Test Practice: SE page 81; Transparencies

Notes ______________________________

____

____

Copyright © McDougal Littell Inc. All rights reserved.

LESSON 2.4

NAME ______________________________ DATE __________

Available as a transparency

WARM-UP EXERCISES

For use before Lesson 2.4, pages 74–81

Tell whether the numbered angles are *adjacent* or *nonadjacent*.

1.

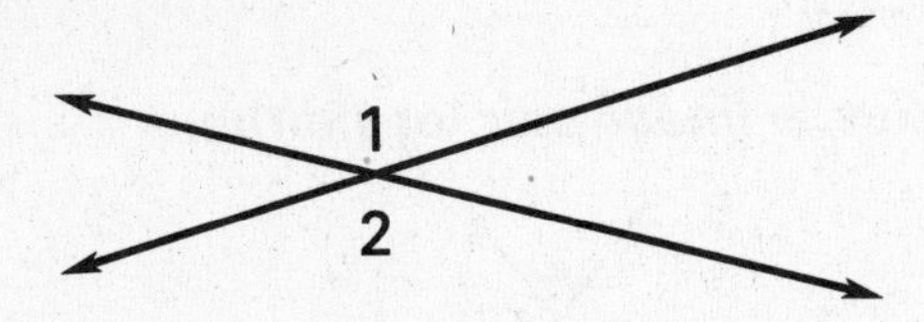

2.

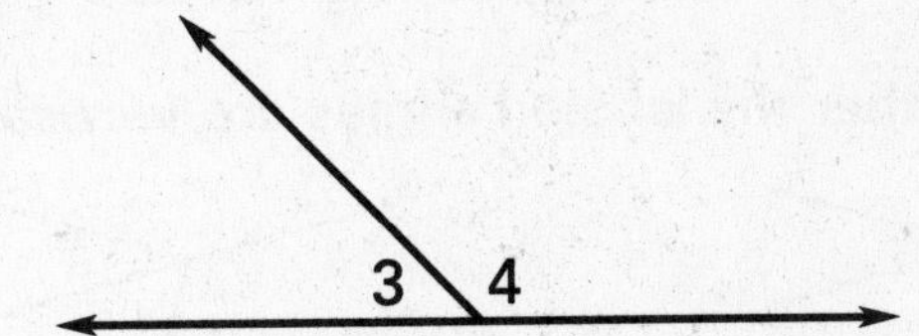

Find the measure of a supplement of the angle given.

3. $m\angle A = 67°$

4. $m\angle B = 148°$

DAILY HOMEWORK QUIZ

For use after Lesson 2.3, pages 67–73

Determine whether the angles are *complementary, supplementary,* or *neither*. Also tell whether the angles are *adjacent* or *nonadjacent*.

1.

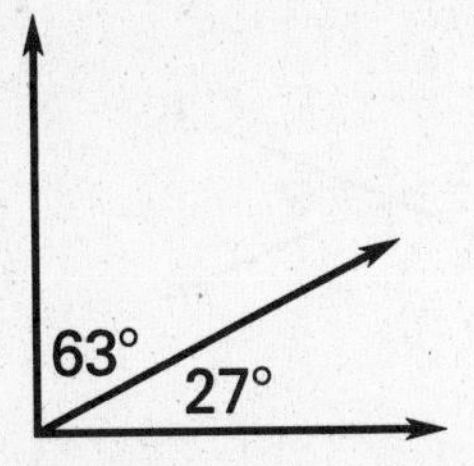

2.

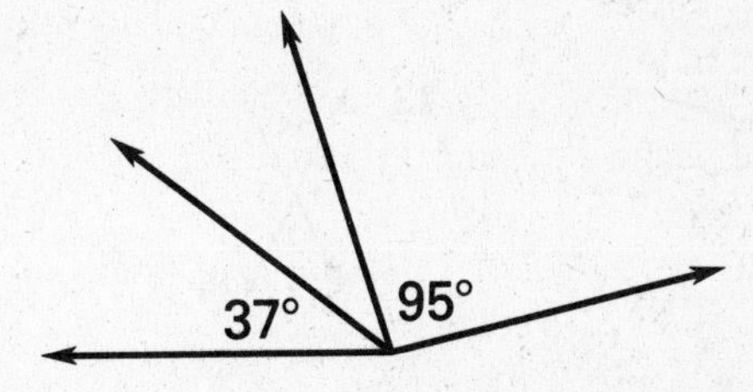

Find the measures of a complement and a supplement of the angle.

3. $m\angle R = 27°$

4. $m\angle T = 11°$

5. In the figure at the right, $\angle ABD$ and $\angle DBC$ are complementary angles. Find the value of x.

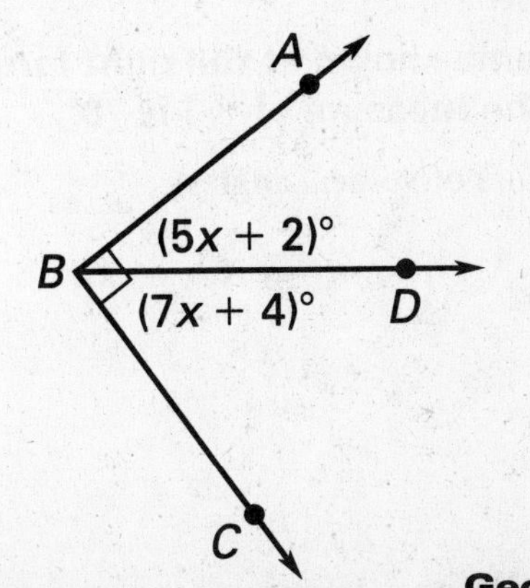

Lesson 2.4

Copyright © McDougal Littell Inc.
All rights reserved.

LESSON 2.4

NAME _______________________________________ DATE ____________

Practice A

For use with pages 74–81

Complete the statement.

1. If two angles form a linear pair, then they are __?__.
2. The Vertical Angles Theorem states that vertical angles are __?__.

Determine whether the labeled angles are *vertical angles,* a *linear pair,* or *neither.*

3. 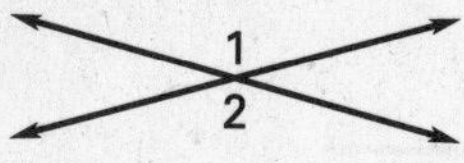

4. 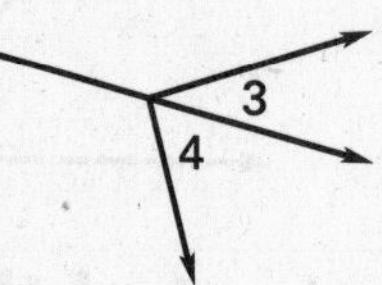

5. 

Use the Linear Pair Postulate to find the measure of ∠2.

6. 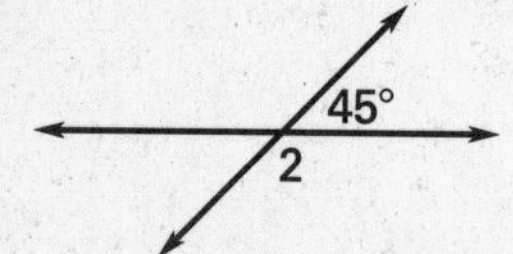

7. 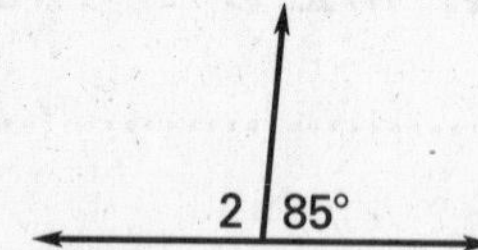

8. 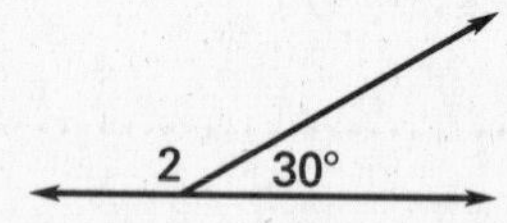

Use the Vertical Angles Theorem to find an angle that is congruent to ∠1.

9. 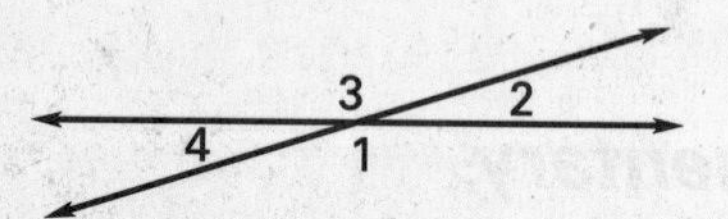

10. 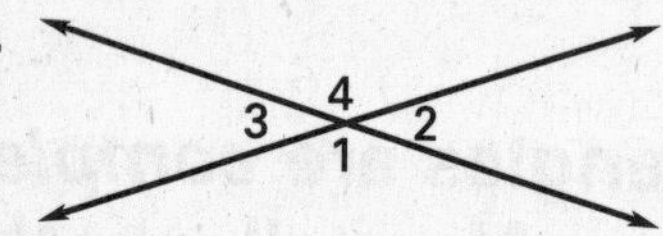

11. 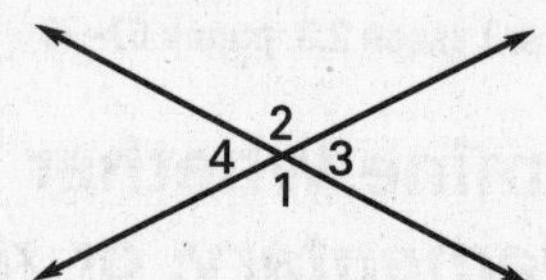

Use the Vertical Angles Theorem and the Linear Pair Postulate to find m∠1, m∠2, and m∠3.

12. 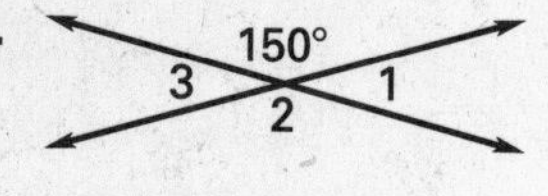

13. 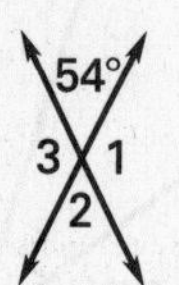

14. 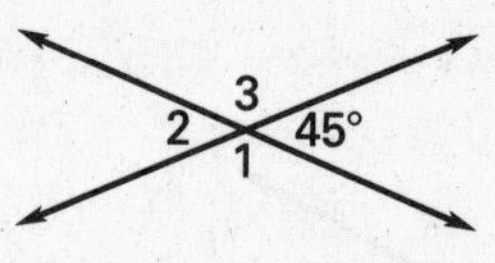

Find the value of *x*.

15. 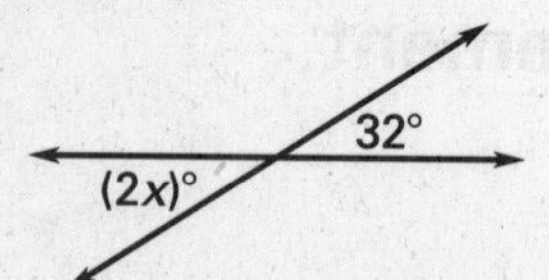

16. 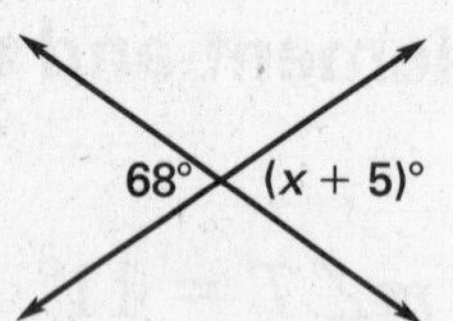

17. 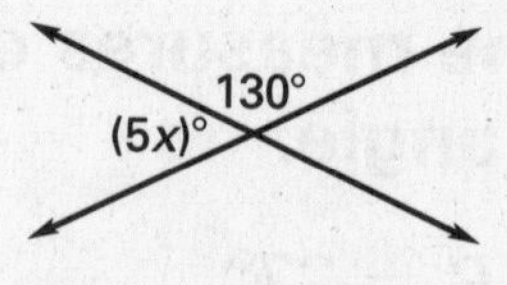

The window frame shown at the right forms angles 1, 2, 3, and 4. The measure of ∠1 is 70°.

18. Name two pairs of vertical angles.
19. Find $m\angle 2$.
20. Find $m\angle 3$.
21. Find $m\angle 4$.

Lesson 2.4

Copyright © McDougal Littell Inc.
All rights reserved.

NAME ____________________ DATE __________

Practice B

For use with pages 74–81

Use the figure at the right to complete the statement.

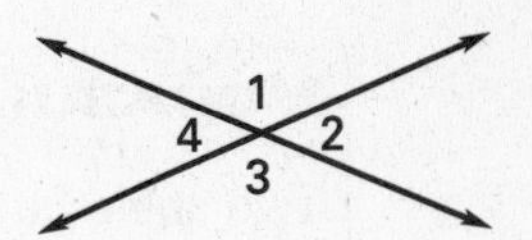

1. $\angle 1$ and __?__ are a linear pair, and so are $\angle 1$ and __?__.
2. $\angle 2$ and __?__ are vertical angles.
3. If $m\angle 3 = 150°$, then $m\angle 2 =$ __?__.
4. $\angle 4 \cong$ __?__.

Find the measure of the numbered angle.

5. 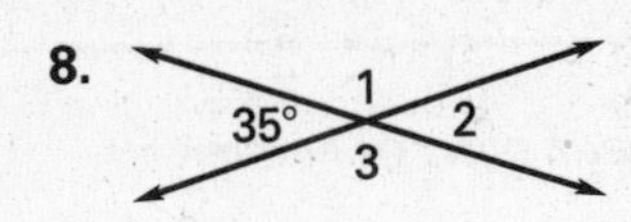

6. 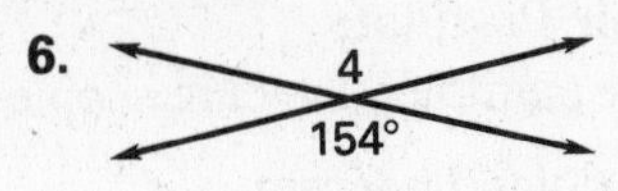

7.

Find $m\angle 1$, $m\angle 2$, and $m\angle 3$.

8.

9.

10.

Use the diagram to complete the statement.

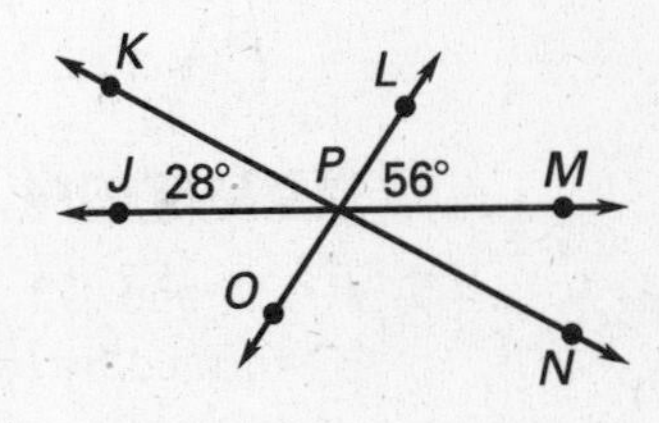

11. $m\angle KPL =$ __?__ °
12. $m\angle LPN =$ __?__ °
13. $m\angle MPN =$ __?__ °
14. $m\angle MPO =$ __?__ °

Find the value of the variable. Then use substitution to find $m\angle PQR$.

15.

16.

17. 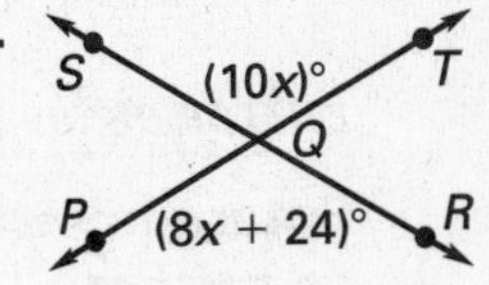

18. The United Kingdom flag can be represented by four intersecting lines that form eight angles. The horizontal and vertical lines are angle bisectors, and the measure of $\angle 1$ is 26.6°. Find the measures of the remaining angles.

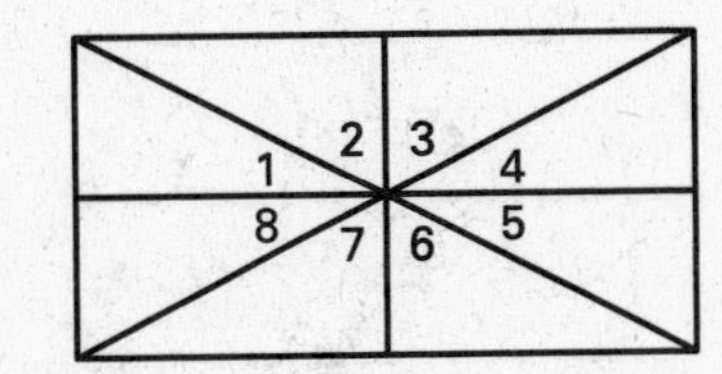

Copyright © McDougal Littell Inc.
All rights reserved.

LESSON 2.4

NAME ______________________________ DATE __________

Reteaching with Practice

For use with pages 74–81

GOAL **Find the measures of angles formed by intersecting lines.**

VOCABULARY

Two angles are **vertical angles** if they are not adjacent and their sides are formed by two intersecting lines.

Two adjacent angles are a **linear pair** if their noncommon sides are on the same line.

Postulate 7 Linear Pair Postulate
If two angles form a linear pair, then they are supplementary.

Theorem 2.3 Vertical Angles Theorem
Vertical angles are congruent.

EXAMPLE 1 ***Identify Vertical Angles and Linear Pairs***

Determine whether the labeled angles are *vertical angles*, a *linear pair*, or *neither*.

a.

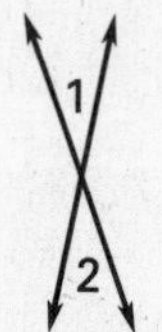

b.

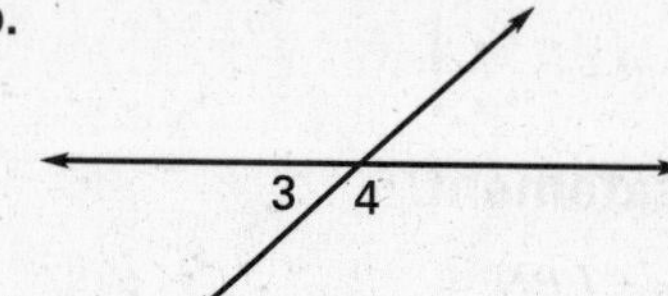

c.

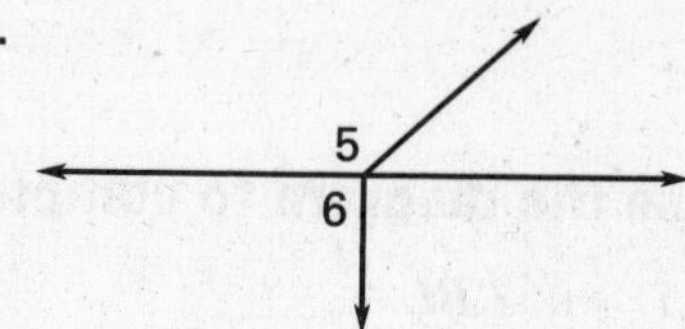

SOLUTION

a. $\angle 1$ and $\angle 2$ are vertical angles because they are not adjacent and their sides are formed by two intersecting lines.

b. $\angle 3$ and $\angle 4$ are a linear pair because they are adjacent and their noncommon sides are on the same line.

c. $\angle 5$ and $\angle 6$ are neither vertical angles nor a linear pair.

Exercises for Example 1

Determine whether the labeled angles are *vertical angles*, a *linear pair*, or *neither*.

1. **2.**

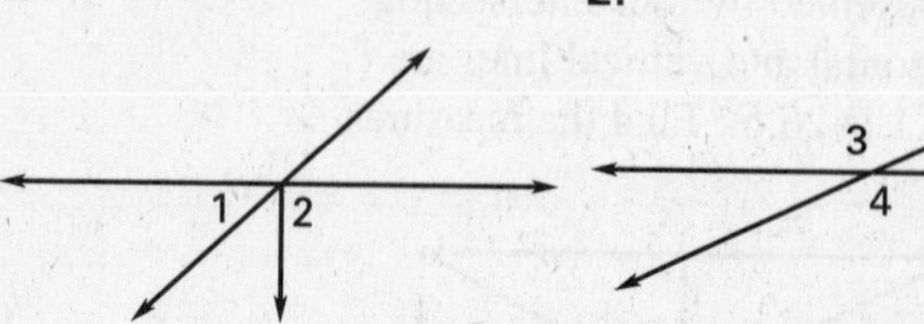

3.

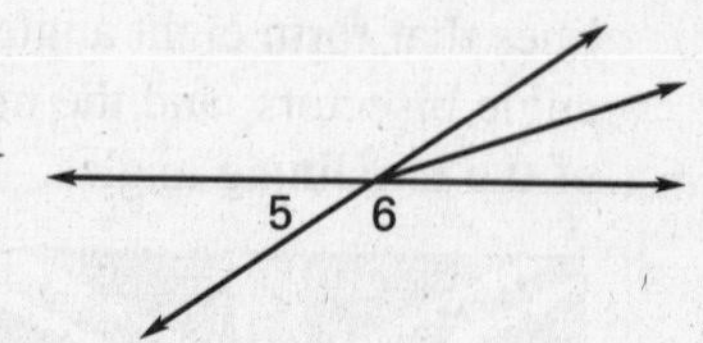

Lesson 2.4

Copyright © McDougal Littell Inc.
All rights reserved.

NAME ______________________________ DATE __________

Reteaching with Practice

For use with pages 74–81

EXAMPLE 2 Use the Linear Pair Postulate

Find the measure of $\angle 1$.

148°

1

SOLUTION

The angles are a linear pair. By the Linear Pair Postulate, they are supplementary.

$m\angle 1 + 148° = 180°$	Definition of supplementary angles.
$m\angle 1 = 32°$	Subtract 148° from each side.

Exercises for Example 2

Find the value of *x*.

4.

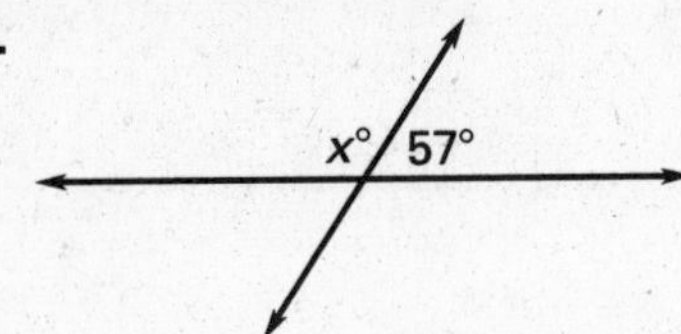

5.

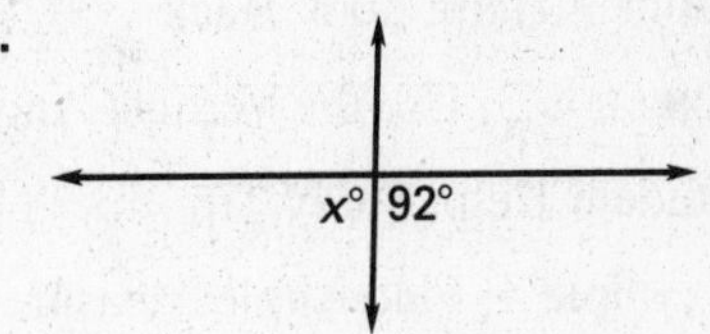

6.

20°

(3x − 5)°

EXAMPLE 3 Use the Vertical Angles Theorem

Find the measure of $\angle ABE$.

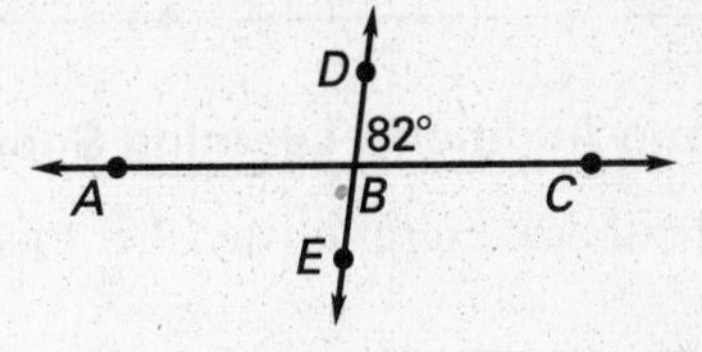

SOLUTION

$\angle DBC$ and $\angle ABE$ are vertical angles. By the Vertical Angles Theorem, $\angle DBC \cong \angle ABE$, so $m\angle DBC = m\angle ABE = 82°$.

Exercises for Example 3

Find the value of the variable.

7.

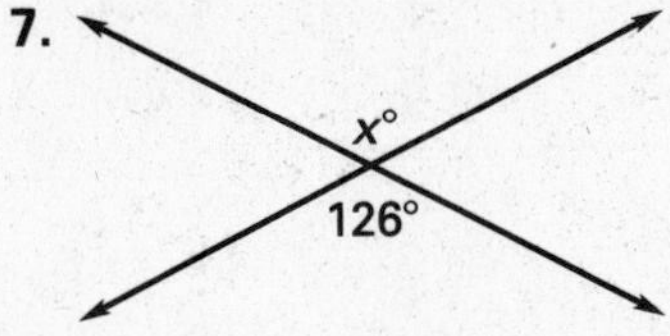

8.

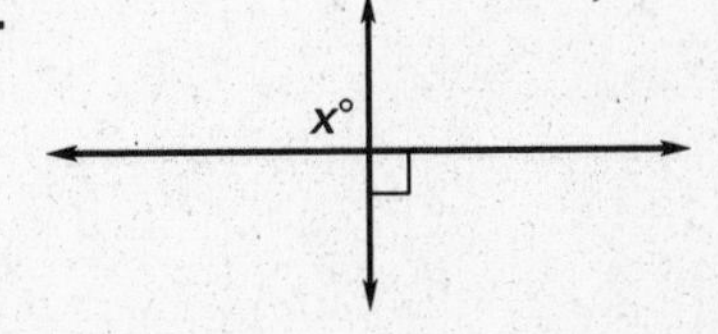

9.

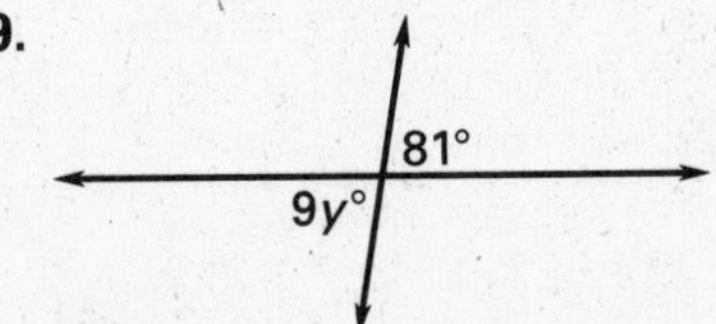

10.

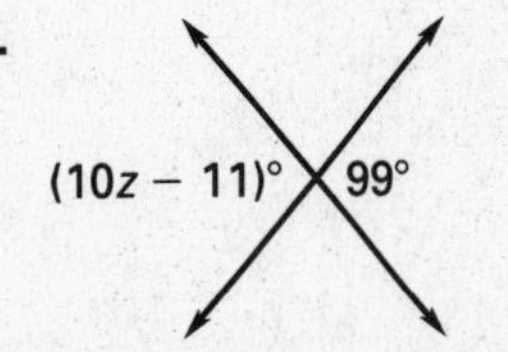

Copyright © McDougal Littell Inc.
All rights reserved.

LESSON 2.4

NAME ______________________________ DATE __________

Quick Catch-Up for Absent Students

For use with pages 74–81

The items checked below were covered in class on (date missed) __________

Activity 2.4: Angles and Intersecting Lines (p. 74)

____ **Goal:** Find the relationship between angles formed by two intersecting lines.

____ Student Help: Vocabulary Tip

Lesson 2.4: Vertical Angles (pp. 75–77)

____ **Goal:** Find the measures of angles formed by intersecting lines.

Material Covered:

____ Example 1: Identify Vertical Angles and Linear Pairs

____ Example 2: Use the Linear Pair Postulate

____ Student Help: Visual Strategy

____ Student Help: Look Back

____ Example 3: Use the Vertical Angles Theorem

____ Student Help: Study Tip

____ Example 4: Find Angle Measures

____ Example 5: Use Algebra with Vertical Angles

Vocabulary:

vertical angles, p. 75 — linear pair, p. 75

____ Other (specify) ______________________________

____

Homework and Additional Learning Support

____ Textbook exercises (teacher to specify) pp. 78–81

____

____ Internet: Homework Help at classzone.com

____ *Reteaching with Practice* worksheet

Lesson 2.4

Copyright © McDougal Littell Inc.
All rights reserved.

LESSON 2.5

TEACHER'S NAME ______________ CLASS ______ ROOM ______ DATE ______

Lesson Plan

2-day lesson (See *Pacing the Chapter,* TE page 50A)

For use with pages 82–87

GOAL Use if-then statements. Apply laws of logic.

State/Local Objectives ______________________________

____

✓ Check the items you wish to use for this lesson.

STARTING OPTIONS

____ Homework Check (2.4): TE page 78; Answer Transparencies
____ Homework Quiz (2.4): TE page 81, CRB page 47, or Transparencies
____ Warm-Up: CRB page 47 or Transparencies

TEACHING OPTIONS

____ Examples: Day 1: 1–2, SE page 82; Day 2: 3–5, SE pages 83–84
____ Extra Examples: TE pages 83–84; Internet Help at classzone.com
____ Checkpoint Exercises: Day 1: 1–2, SE page 83; Day 2: 3–6, SE pages 83–84
____ Concept Check: TE page 84
____ Guided Practice Exercises: Day 1: 1–4, SE page 85; Day 2: 5–6, SE page 85

APPLY/HOMEWORK

Homework Assignment

____ Basic: Day 1: pp. 85–87 Exs. 7–12, 16, 20, 24–42 even
Day 2: pp. 85–87 Exs. 13–19 odd, 23–43 odd
____ Average: Day 1: pp. 85–87 Exs. 7–12, 20, 21, 24–42 even
Day 2: pp. 85–87 Exs. 13–19, 22, 23–43 odd
____ Advanced: Day 1: pp. 85–87 Exs. 8–12 even, 20, 21, 24–34
Day 2: pp. 85–87 Exs. 14–18 even, 22, 23, 35–43; EC: TE p. 50D*, classzone.com

Reteaching the Lesson

____ Practice Masters: CRB pages 48–49 (Level A, Level B)
____ Reteaching with Practice: CRB pages 50–51 or Practice Workbook with Examples;
Resources in Spanish

Extending the Lesson

____ Challenge: TE page 50D; classzone.com

ASSESSMENT OPTIONS

____ Daily Quiz (2.5): TE page 87, CRB page 55, or Transparencies
____ Standardized Test Practice: SE page 87; Transparencies

Notes ______________________________

____

____

Copyright © McDougal Littell Inc.
All rights reserved.

TEACHER'S NAME ______________________ CLASS ______ ROOM ______ DATE ______

Lesson Plan for Block Scheduling

1-block lesson (See *Pacing the Chapter,* TE page 50A)

For use with pages 82–87

GOAL Use if-then statements. Apply laws of logic.

State/Local Objectives ______________________

CHAPTER PACING GUIDE	
Day	**Lesson**
1	2.1
2	2.2
3	2.3
4	2.4
5	**2.5**
6	2.6
7	Ch. 2 Review and Assess

✓ **Check the items you wish to use for this lesson.**

STARTING OPTIONS

____ Homework Check (2.4): TE page 78; Answer Transparencies
____ Homework Quiz (2.4): TE page 81, CRB page 47, or Transparencies
____ Warm-Up: CRB page 47 or Transparencies

TEACHING OPTIONS

____ Examples: 1–5, SE pages 82–84
____ Extra Examples: TE pages 83–84; Internet Help at classzone.com
____ Checkpoint Exercises: 1–6, SE pages 83–84
____ Concept Check: TE page 84
____ Guided Practice Exercises: 1–6, SE page 85

APPLY/HOMEWORK

Homework Assignment

____ Block Schedule: pp. 85–87 Exs. 7–43

Reteaching the Lesson

____ Practice Masters: CRB pages 48–49 (Level A, Level B)
____ Reteaching with Practice: CRB pages 50–51 or Practice Workbook with Examples; Resources in Spanish

Extending the Lesson

____ Challenge: TE page 50D; classzone.com

ASSESSMENT OPTIONS

____ Daily Quiz: (2.5): TE page 87, CRB page 55, or Transparencies
____ Standardized Test Practice: SE page 87; Transparencies

Notes ______________________

Copyright © McDougal Littell Inc. All rights reserved.

LESSON **2.5** NAME ______________________ DATE ________

WARM-UP EXERCISES

For use before Lesson 2.5, pages 82–87

Determine whether the statement is *true* or *false*.

1. $\angle C$ and $\angle D$ are vertical angles, so they are congruent.

2. $\angle A$ and $\angle B$ are supplementary angles, so they form a linear pair.

3. $\angle ABC$ is bisected by $\overrightarrow{BD}$, so $m\angle ABC = \frac{1}{2}m\angle ABD$.

DAILY HOMEWORK QUIZ

For use after Lesson 2.4, pages 74–81

Use the diagram at the right.

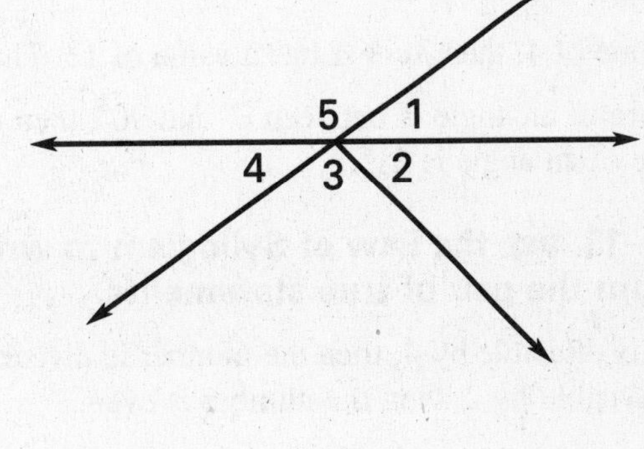

1. Name a linear pair.

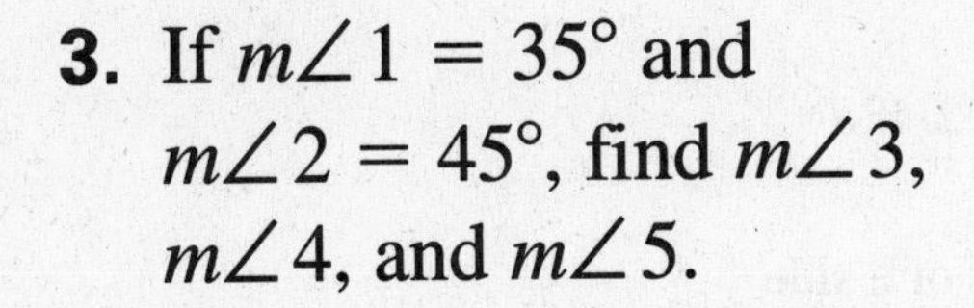

2. Name a pair of vertical angles.

3. If $m\angle 1 = 35°$ and $m\angle 2 = 45°$, find $m\angle 3$, $m\angle 4$, and $m\angle 5$.

Use the diagram at the right to find the value of each variable.

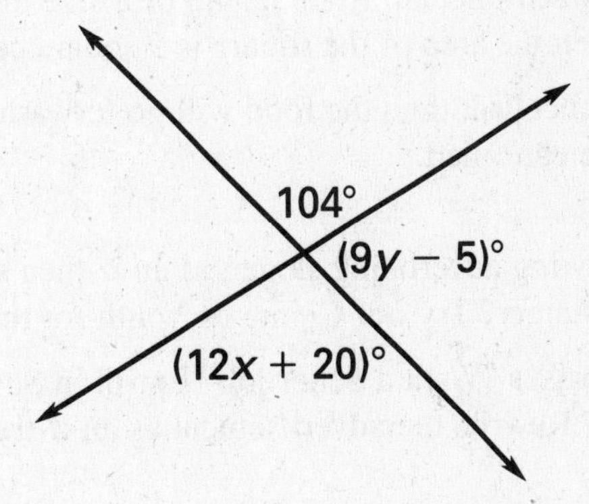

4. x

5. y

Copyright © McDougal Littell Inc.
All rights reserved.

LESSON 2.5

NAME ________________________________ DATE ____________

Practice A

For use with pages 82–87

Identify the hypothesis and the conclusion of the if-then statement.

1. If two angles have the same measure, then the angles are congruent.
2. If two angles form a linear pair, then the angles are supplementary.
3. If the sum of the measures of two angles is 90°, then the angles are complementary.
4. If the measure of an angle is 90°, then the angle is a right angle.

Rewrite the statement as an if-then statement.

5. I will purchase a school yearbook if it costs less than $20.
6. Your team will travel to the state championship game if it wins the district championship.
7. You cannot ride your bicycle if it has a flat tire.
8. School will be cancelled if it snows six inches overnight.

Using the Law of Detachment, what can you conclude from the true statements?

9. If x has a value of 4, then $3x + 1$ has a value of 13. The value of x is 4.
10. If the measure of an angle is between 0° and 90°, then the angle is acute. The measure of an angle is 51°.

In Exercises 11–13, use the Law of Syllogism to write the statement that follows from the pair of true statements.

11. If a number is divisible by 4, then the number is divisible by 2. If a number is divisible by 2, then the number is even.
12. If the perimeter of a square is 12 centimeters, then the length of a side of the square is 3 centimeters. If the length of a side of a square is 3 centimeters, then the area of the square is 9 square centimeters.
13. If the picnic is cancelled, then the food will go to waste. If it rains, then the picnic will be cancelled.

14. Rewrite the following advertising slogan as an if-then statement: "Want to look younger? Try our Creme de Youth for thirty days."
15. A billboard advertises "Want a better job? Enroll in Seville College's on-line degree program." Rewrite the advertisement as an if-then statement.

Copyright © McDougal Littell Inc.
All rights reserved.

LESSON 2.5

NAME ______________________ DATE ________

Practice B

For use with pages 82–87

Write the hypothesis and the conclusion of the if-then statement.

1. If two planes intersect, then their intersection is a line.
2. If $\angle A$ is acute, then the measure of $\angle A$ is between 0° and 90°.
3. If the sum of the measures of two angles is 180°, then the angles are supplementary.
4. If the measure of an angle is between 90° and 180°, then the angle is obtuse.

Rewrite the statement as an if-then statement.

5. Two angles that have the same measure are congruent angles.
6. Two angles that form a linear pair are supplementary angles.
7. An angle that has a measure of 90° is a right angle.
8. An angle that has a measure between 90° and 180° is an obtuse angle.
9. A dog with proper training will not misbehave.

What law of logic is illustrated in the following statements? What can you conclude if the statements are true?

10. If you earn more than $14, you can buy a new CD. You earn $15.
11. If the area of a square is 49 square inches, then the length of a side of the square is 7 inches. If the length of a side of a square is 7 inches, then the perimeter of the square is 28 inches.

In Exercises 12 and 13, write the statement that follows from the pair of true statements.

12. If the width of a rectangle is 5 centimeters and the length of the rectangle is 2 centimeters, then the area of the rectangle is 10 square centimeters. The width of a rectangle is 5 centimeters and the length of the rectangle is 2 centimeters.
13. If a number is divisible by 10, then the number is divisible by 2. If a number is divisible by 2, then the number is even.

14. Rewrite the following advertising slogan as an if-then statement: "Want to learn about computers? Try our Computer Wizard tutorial for thirty days."
15. Identify the hypothesis and the conclusion of the if-then statement in Exercise 14.

Copyright © McDougal Littell Inc.
All rights reserved.

LESSON 2.5

NAME ______________________________ DATE ____________

Reteaching with Practice

For use with pages 82–87

GOAL Use if-then statements. Apply laws of logic.

VOCABULARY

An **if-then statement** has two parts. The "if" part contains the **hypothesis.** The "then" part contains the **conclusion.**

Deductive reasoning uses facts, definitions, accepted properties, and the laws of logic to make a logical argument.

Laws of Logic

Law of Detachment

If the hypothesis of a true if-then statement is true, then the conclusion is also true.

Law of Syllogism

If *statement p*, then *statement q*.
If *statement q*, then *statement r*. → If these statements are true,
If *statement p*, then *statement r*. ← then this statement is true.

EXAMPLE 1 Identify the Hypothesis and Conclusion

Identify the hypothesis and the conclusion of the if-then statement.

If I do my chores, then I will be given my allowance.

SOLUTION

"I do my chores" is the hypothesis.

"I will be given my allowance" is the conclusion.

Exercises for Example 1

Identify the hypothesis and the conclusion of the statement.

1. If it is raining outside, then there are clouds in the sky.

2. If two angles are vertical angles, then the two angles are congruent.

EXAMPLE 2 Write If-Then Statements

Rewrite the statement as an if-then statement.

a. I will get a B in History class if I get an A on the final exam in that class.

b. A right angle measures 90°.

SOLUTION

a. If I get an A on the final exam in History class, then I will get a B in that class.

b. If an angle is a right angle, then the angle measures 90°.

Copyright © McDougal Littell Inc.
All rights reserved.

NAME ______________________________ DATE __________

Reteaching with Practice

For use with pages 82–87

Exercises for Example 2

Rewrite the statement as an if-then statement.

3. An even number is divisible by two.

4. The measures of two angles are equal if the two angles are congruent.

EXAMPLE 3 Use the Law of Detachment

Which argument is correct?

Argument 1: If two angles measure 115° and 65°, then the angles are supplementary. $\angle 1$ and $\angle 2$ are supplementary. So, $m\angle 1 = 115°$ and $m\angle 2 = 65°$.

Argument 2: If two angles measure 115° and 65°, then the angles are supplementary. The measure of $\angle 1$ is 115° and the measure of $\angle 2$ is 65°. So, $\angle 1$ and $\angle 2$ are supplementary.

SOLUTION

Argument 2 is correct. The hypothesis (two angles measure 115° and 65°) is true, which implies that the conclusion (they are supplementary) is true by the Law of Detachment.

Exercise for Example 3

Determine which argument is correct. Explain your reasoning.

5. *Argument 1:* If it is noon on Monday, then the children are in school. The children are in school. So, it is noon on Monday.

   Argument 2: If it is noon on Monday, then the children are in school. It is noon on Monday. So, the children are in school.

EXAMPLE 4 Use the Law of Syllogism

Write the statement that follows from the pair of true statements.

If the juice is knocked over, then it will spill on the carpet.
If the juice spills on the carpet, then it will stain the carpet.

SOLUTION

Use the Law of Syllogism. If the juice is knocked over, then it will stain the carpet.

Exercise for Example 4

Write the statement that follows from the pair of true statements.

6. If I throw the stick, then my dog will go fetch it.
   If my dog fetches the stick, then my dog will bring it back to me.

Copyright © McDougal Littell Inc.
All rights reserved.

LESSON 2.5

NAME ______________________________ DATE __________

Quick Catch-Up for Absent Students

For use with pages 82–87

The items checked below were covered in class on (date missed) __________

Lesson 2.5: If-Then Statements and Deductive Reasoning (pp. 82–84)

____ **Goal:** Use if-then statements. Apply laws of logic.

Material Covered:

____ Student Help: Vocabulary Tip

____ Example 1: Identify the Hypothesis and Conclusion

____ Student Help: Study Tip

____ Example 2: Write If-Then Statements

____ Example 3: Use the Law of Detachment

____ Example 4: Use the Law of Detachment

____ Example 5: Use the Law of Syllogism

Vocabulary:

if-then statement, p. 82
hypothesis, p. 82
conclusion, p. 82
deductive reasoning, p. 83

____ Other (specify) ______________________________

____

Homework and Additional Learning Support

____ Textbook exercises (teacher to specify) pp. 85–87

____

____ Internet: More Examples at classzone.com

____ *Reteaching with Practice* worksheet

Copyright © McDougal Littell Inc.
All rights reserved.

LESSON 2.6

TEACHER'S NAME ____________________ CLASS ______ ROOM ______ DATE ______

Lesson Plan

2-day lesson (See *Pacing the Chapter,* TE page 50A)

For use with pages 88–94

GOAL **Use properties of equality and congruence.**

State/Local Objectives ____________________

____

✓ Check the items you wish to use for this lesson.

STARTING OPTIONS

____ Homework Check (2.5): TE page 85; Answer Transparencies
____ Homework Quiz (2.5): TE page 87, CRB page 55, or Transparencies
____ Warm-Up: CRB page 55 or Transparencies

TEACHING OPTIONS

____ Examples: Day 1: 1, SE page 89; Day 2: 2–3, SE pages 89–90
____ Extra Examples: TE pages 89–90; Internet Help at classzone.com
____ Checkpoint Exercises: Day 1: 1–3, SE page 89; Day 2: 4–5, SE pages 89–90
____ Technology Activity with Keystrokes: CRB pages 56–58
____ Concept Check: TE page 90
____ Guided Practice Exercises: Day 1: 1–9, SE page 91

APPLY/HOMEWORK

Homework Assignment

____ Basic: Day 1: pp. 91–94 Exs. 10–18, 28, 30–33, 36–43
Day 2: pp. 91–94 Exs. 19–21, 29, 34, 35, Quiz 2
____ Average: Day 1: pp. 91–94 Exs. 10–18, 28, 30–43
Day 2: pp. 91–94 Exs. 19–23, 25, 26, 29, Quiz 2
____ Advanced: Day 1: pp. 91–94 Exs. 13–18, 28, 30–43
Day 2: pp. 91–94 Exs. 19–26, 27*, 29, Quiz 2; EC: classzone.com

Reteaching the Lesson

____ Practice Masters: CRB pages 59–60 (Level A, Level B)
____ Reteaching with Practice: CRB pages 61–62 or Practice Workbook with Examples; Resources in Spanish

Extending the Lesson

____ Challenge: SE page 93; classzone.com

ASSESSMENT OPTIONS

____ Daily Quiz (2.6): TE page 94 or Transparencies
____ Standardized Test Practice: SE page 93; Transparencies
____ Quiz (2.4–2.6): SE page 94; CRB page 64

Notes ____________________

____

____

Copyright © McDougal Littell Inc.
All rights reserved.

LESSON 2.6

TEACHER'S NAME ______________ CLASS ______ ROOM ______ DATE ______

Lesson Plan for Block Scheduling

1-block lesson (See *Pacing the Chapter,* TE page 50A) For use with pages 88–94

GOAL Use properties of equality and congruence.

State/Local Objectives ____________________

____

____

CHAPTER PACING GUIDE	
Day	**Lesson**
1	2.1
2	2.2
3	2.3
4	2.4
5	2.5
6	**2.6**
7	Ch. 2 Review and Assess

✓ Check the items you wish to use for this lesson.

STARTING OPTIONS

____ Homework Check (2.5): TE page 85; Answer Transparencies
____ Homework Quiz (2.5):TE page 87, CRB page 55, or Transparencies
____ Warm–Up: CRB page 55 or Transparencies.

TEACHING OPTIONS

____ Examples: 1–3, SE pages 89–90
____ Extra Examples: TE pages 89–90; Internet Help at classzone.com
____ Checkpoint Exercises: 1–5, SE pages 89–90
____ Technology Activity with Keystrokes: CRB pages 56–58
____ Concept Check: TE page 90
____ Guided Practice Exercises: 1–9, SE page 91

APPLY/HOMEWORK

Homework Assignment

____ Block Schedule: pp. 91–94 Exs. 10–23, 25, 26, 28–43, Quiz 2

Reteaching the Lesson

____ Practice Masters: CRB pages 59–60 (Level A, Level B)
____ Reteaching with Practice: CRB pages 61–62 or Practice Workbook with Examples; Resources in Spanish

Extending the Lesson

____ Challenge: SE page 93; classzone.com

ASSESSMENT OPTIONS

____ Daily Quiz (2.6): TE page 94 or Transparencies
____ Standardized Test Practice: SE page 93; Transparencies
____ Quiz (2.4–2.6): SE page 94; CRB page 64

Notes ____________________

____

____

Copyright © McDougal Littell Inc.
All rights reserved.

LESSON 2.6

NAME ______________________________ DATE __________

Available as a transparency

WARM-UP EXERCISES

For use before Lesson 2.6, pages 88–94

Solve the equation.

1. $3x = 27$
2. $x + 6 = -17$
3. $x - 9 = 20$
4. $\frac{2}{3}x = 6$
5. $-x = 4$

DAILY HOMEWORK QUIZ

For use after Lesson 2.5, pages 82–87

1. Identify the hypothesis and the conclusion of the if-then statement.

   If the temperature begins falling, then the rain will change to snow.

2. Rewrite the statement as an if-then statement.

   The measure of a right angle is 90°.

3. What can you conclude from the given true statements?

   If two lines are parallel, then they never meet. Two lines are parallel.

4. Write the if-then statement that follows from the pair of true statements.

   If the rain becomes sleet, the roads will ice over. If the roads ice over, then school will be delayed.

Copyright © McDougal Littell Inc.
All rights reserved.

LESSON **2.6**

NAME ______________________________ DATE ____________

Technology Activity

For use with pages 88–94

GOAL **Use geometry software to verify a mathematical statement.**

Geometry software can be used to show that a mathematical statement is true. For example, you could use geometry software to construct the diagram below. Then, you could use the software's measuring tool to verify the statement about the segment length.

GIVEN: $AC = DB$

SHOW: $AD = CB$

Activity

1. Construct a diagram with points A, B, C, and D such that $AC = DB$ (see figure above).
2. Measure the lengths of $\overline{AC}$, $\overline{DB}$, $\overline{AD}$, and $\overline{CB}$ to verify that $AD = CB$.

Exercises

Use geometry software to verify the following.

1. GIVEN: $\overline{AD} \cong \overline{BC}$, $\overline{EC} \cong \overline{ED}$

SHOW: $\overline{AE} \cong \overline{BE}$

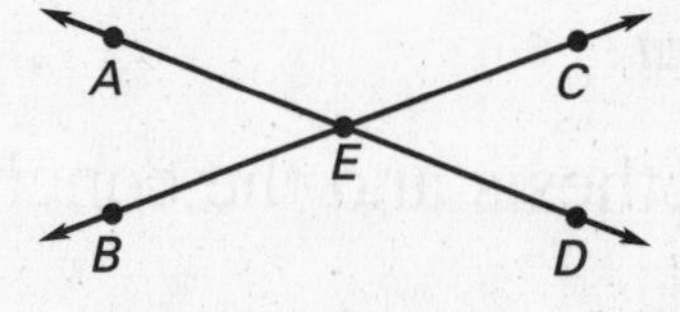

2. GIVEN: $AB = BC$

SHOW: $AC = 2 \cdot BC$

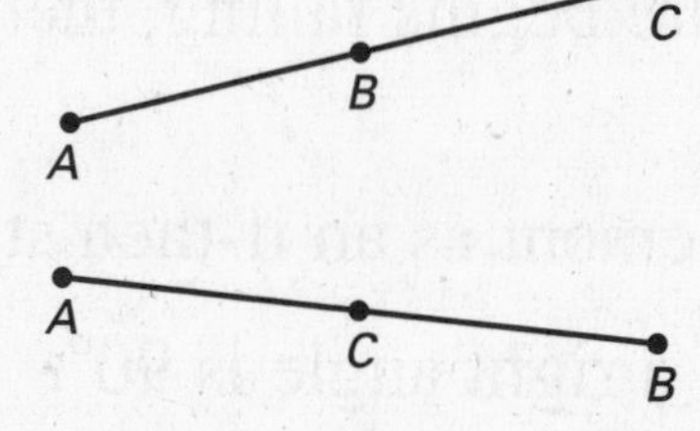

3. GIVEN: C is the midpoint of $\overline{AB}$.

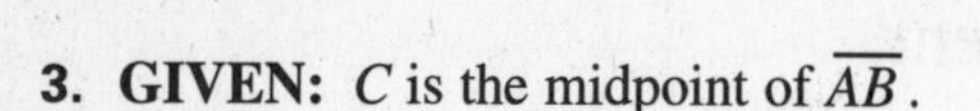

SHOW: $AC = \frac{1}{2}AB$ and $CB = \frac{1}{2}AB$

Copyright © McDougal Littell Inc.
All rights reserved.

Lesson 2.6

NAME _______________ DATE _______

Technology Keystrokes

For use with Technology Activity, CRB page 56

TI-92

1. Construct a segment with points A, B, C, and D so that $AC = DB$.

   F2 5 (Move cursor to desired location.) ENTER A (Move cursor to desired location.) ENTER B

   F2 2 (Move cursor to desired location on $\overline{AB}$.) ENTER C (Move cursor to desired location.) ENTER D

   F6 1 (Move cursor to point C.) ENTER (Move cursor to point A.) ENTER (Move cursor to point D.) ENTER (Move cursor to point B.) ENTER

   F1 1 (Move cursor to point D.) Use the drag key [drag key] and the cursor pad to drag point D along $\overline{AB}$ until $AC = BD$.

2. Measure AD and CB.

   F6 1 (Move cursor to point A.) ENTER (Move cursor to point D.) ENTER (Move cursor to point C.) ENTER (Move cursor to point B.) ENTER

Copyright © McDougal Littell Inc.
All rights reserved.

LESSON 2.6 CONTINUED

NAME ______________________ DATE ____________

Technology Keystrokes

For use with Technology Activity, CRB page 56

SKETCHPAD

1. Construct a segment with points *A, B, C,* and *D*.
   Choose the segment tool. Draw $\overline{AB}$. Choose the point tool. Draw points *C* and *D* on $\overline{AB}$. Make sure that your diagram looks similar to the one given.
2. Find *AC*, *DB*, *AD*, and *CB*.
   Choose the selection arrow tool. Measure the length of $\overline{AC}$. Select point *A*. Hold down the shift key and select point *C*. Choose **Distance** from the **Measure** menu. Repeat this procedure to measure the lengths of $\overline{DB}$, $\overline{AD}$, and $\overline{CB}$.
3. Locate the points such that $AC = DB$.
   Select any of the points and drag them until $AC = DB$.

Lesson 2.6

Copyright © McDougal Littell Inc.
All rights reserved.

LESSON **2.6**

NAME ______________________ DATE ________

Practice A

For use with pages 88–94

Match the statement with the property it illustrates.

1. $\angle B \cong \angle B$
2. If $\overline{PQ} \cong \overline{RS}$, then $\overline{RS} \cong \overline{PQ}$.
3. If $m\angle A = m\angle B$ and $m\angle B = m\angle C$, then $m\angle A = m\angle C$.
4. If $\overline{MN} \cong \overline{OP}$ and $\overline{OP} \cong \overline{QR}$, then $\overline{MN} \cong \overline{QR}$.
5. $m\angle 1 = m\angle 1$
6. If $m\angle 3 = m\angle 4$, then $m\angle 4 = m\angle 3$.

A. Reflexive Property of Equality
B. Symmetric Property of Equality
C. Transitive Property of Equality
D. Reflexive Property of Congruence
E. Symmetric Property of Congruence
F. Transitive Property of Congruence

Name the property of equality that the statement illustrates.

7. If $m\angle 1 = m\angle 4$, then $m\angle 1 - 30° = m\angle 4 - 30°$.
8. If $LM = NP$, then $2 \cdot LM = 2 \cdot NP$.
9. If $XY = EF$, then $XY + 7 = EF + 7$.
10. If $m\angle A = m\angle B$, then $\frac{m\angle A}{3} = \frac{m\angle B}{3}$.
11. If $CD = 4$, then $CD + 12 = 4 + 12$.
12. In the diagram, $AB + BC = 12$, and $BC = 3$. Complete the argument to show that $AB = 9$.

A B C

$AB + BC = 12$	Given
$BC = 3$	Given
$AB + 3 = 12$	___?___ Property of Equality
$AB = 9$	___?___ Property of Equality

13. In the figures at the right, $\angle JKL \cong \angle EDF$, and $\angle EDF \cong \angle CDE$. Complete the argument to show that $\angle CDE \cong \angle JKL$.

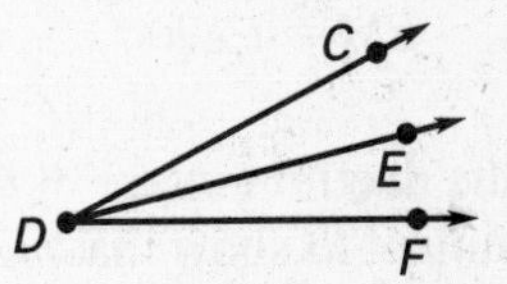

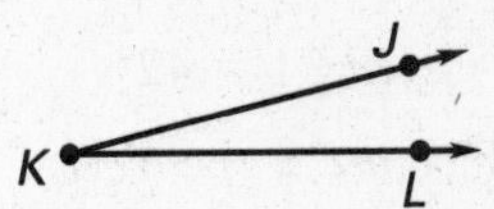

$\angle JKL \cong \angle EDF$	Given
$\angle EDF \cong \angle CDE$	Given
$\angle JKL \cong \angle CDE$	___?___ Property of Congruence
$\angle CDE \cong \angle JKL$	___?___ Property of Congruence

Identify the property of equality that is illustrated.

14.     =   

15. If 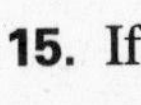    =  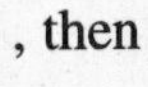, then  =   

Lesson 2.6

Copyright © McDougal Littell Inc.
All rights reserved.

LESSON 2.6

NAME ______________________ DATE ________

Practice B

For use with pages 88–94

Use the property to complete the statement.

1. Reflexive Property of Equality: $m\angle A =$ __?__.
2. Symmetric Property of Equality: If $EF = GH$, then __?__ $=$ __?__.
3. Transitive Property of Equality: If $m\angle 1 = m\angle 2$ and $m\angle 2 = m\angle 3$, then __?__ $=$ __?__.
4. Reflexive Property of Congruence: __?__ $\cong \overline{KL}$
5. Symmetric Property of Congruence: If $\angle 5 \cong \angle 6$, then __?__ $\cong$ __?__.
6. Transitive Property of Congruence: If $\overline{AB} \cong \overline{CD}$ and $\overline{CD} \cong \overline{EF}$, then __?__ $\cong$ __?__.

Name the property that the statement illustrates.

7. If $m\angle 1 = m\angle 2$, then $m\angle 1 - m\angle 3 = m\angle 2 - m\angle 3$.
8. If $KL = PQ$, then $2 \cdot KL = 2 \cdot PQ$.
9. If $m\angle A = m\angle C$, then $m\angle A + m\angle D = m\angle C + m\angle D$.
10. If $xy = 5$, then $\frac{xy}{3} = \frac{5}{3}$.
11. If $m\angle 2 = 90°$ and $m\angle 3 = m\angle 2 + 44°$, then $m\angle 3 = 90° + 44°$.

12. In the diagram at the right, M is the midpoint of $\overline{EF}$. Complete the argument to show that $EM = \frac{1}{2}EF$.

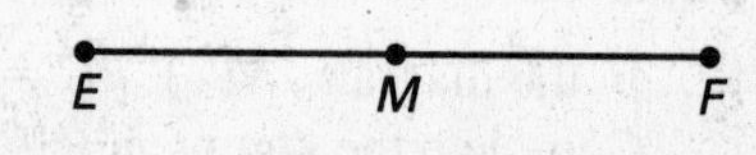

$EM = MF$	Definition of __?__
$EM + MF = EF$	__?__ Postulate
$EM + EM = EF$	__?__ Property of Equality
$2 \cdot EM = EF$	Distributive property
$EM = \frac{1}{2}EF$	__?__ Property of Equality

13. In the diagram, $m\angle 1 + m\angle 2 = 98°$, and $m\angle 1 = 42°$. Complete the argument to show that $m\angle 2 = 56°$.

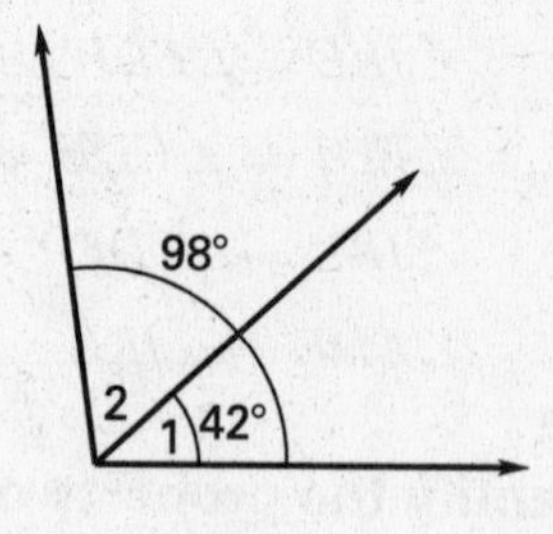

$m\angle 1 + m\angle 2 = 98°$	Given
$m\angle 1 = 42°$	Given
$42° + m\angle 2 = 98°$	__?__
$m\angle 2 = 56°$	__?__

Copyright © McDougal Littell Inc.
All rights reserved.

NAME ______________________ DATE __________

Reteaching with Practice

For use with pages 88–94

GOAL Use properties of equality and congruence.

Vocabulary

Properties of Equality and Congruence

Reflexive Property

Equality: $AB = AB$ **Congruence:** $\overline{AB} \cong \overline{AB}$

$m\angle A = m\angle A$ $\angle A \cong \angle A$

Symmetric Property

Equality: If $AB = CD$, then $CD = AB$. **Congruence:** If $\overline{AB} \cong \overline{CD}$, then $\overline{CD} \cong \overline{AB}$.

If $m\angle A = m\angle B$, then $m\angle B = m\angle A$. If $\angle A \cong \angle B$, then $\angle B \cong \angle A$.

Transitive Property

Equality: If $AB = CD$ and $CD = EF$, then $AB = EF$. **Congruence:** If $\overline{AB} \cong \overline{CD}$ and $\overline{CD} \cong \overline{EF}$, then $\overline{AB} \cong \overline{EF}$.

If $m\angle A = m\angle B$ and $m\angle B = m\angle C$, then $m\angle A = m\angle C$. If $\angle A \cong \angle B$ and $\angle B \cong \angle C$, then $\angle A \cong \angle C$.

Properties of Equality

Addition Property

Adding the same number to each side of an equation produces an equivalent equation.

Subtraction Property

Subtracting the same number from each side of an equation produces an equivalent equation.

Multiplication Property

Multiplying each side of an equation by the same nonzero number produces an equivalent equation.

Division Property

Dividing each side of an equation by the same nonzero number produces an equivalent equation.

Substitution Property

Substituting a number for a variable in an equation produces an equivalent equation.

EXAMPLE 1 ***Name Properties of Equality and Congruence***

Name the property that the statement illustrates.

a. $\overline{ST} \cong \overline{ST}$

b. If $m\angle R = m\angle S$ and $m\angle S = m\angle T$, then $m\angle R = m\angle T$.

c. If $AB = EF$, then $EF = AB$.

Solution

a. Reflexive Property of Congruence

b. Transitive Property of Equality

c. Symmetric Property of Equality

Copyright © McDougal Littell Inc.
All rights reserved.

NAME ______________________ DATE ________

Reteaching with Practice

For use with pages 88–94

Exercises for Example 1

Name the property that the statement illustrates.

1. If $\overline{MN} \cong \overline{OP}$ and $\overline{OP} \cong \overline{QR}$, then $\overline{MN} \cong \overline{QR}$.
2. If $\angle G \cong \angle Z$, then $\angle Z \cong \angle G$.

EXAMPLE 2 Use Properties of Equality

In the diagram, $\angle 1$ and $\angle 2$ are a linear pair and $m\angle 1 = m\angle 3$. Show that $m\angle 2 + m\angle 3 = 180°$.

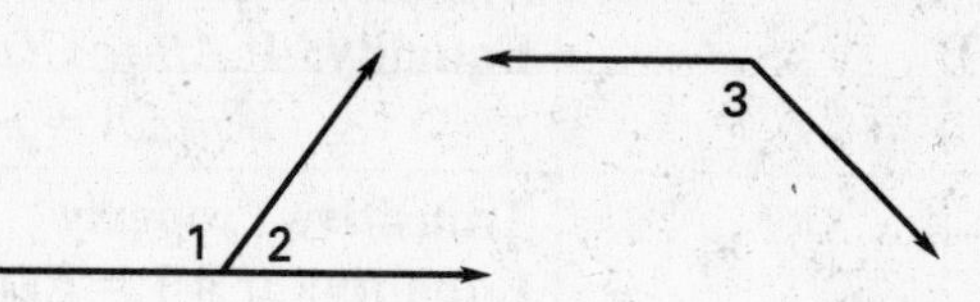

SOLUTION

$m\angle 2 + m\angle 1 = 180°$	Linear Pair Postulate
$m\angle 1 = m\angle 3$	Given
$m\angle 2 + m\angle 3 = 180°$	Substitution Property of Equality

Exercise for Example 2

In the diagram, $\overrightarrow{AC}$ bisects $\angle BAD$ and $\overrightarrow{AD}$ bisects $\angle CAE$. Complete the argument to show that $\angle BAC \cong \angle DAE$.

3. $\angle BAC \cong \angle CAD$ Definition of ________
   $\angle CAD \cong \angle DAE$ Definition of ________
   $\angle BAC \cong \angle DAE$ ________ Property of ________

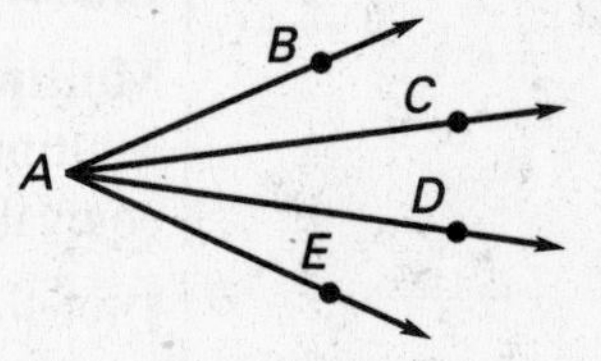

EXAMPLE 3 Use Algebra

Find the value of x, given that $EF = FG$ and $FG = GH$.

SOLUTION

$EF = GH$	Transitive Property of Equality
$4x + 3 = 23$	Substitute $(4x + 3)$ for EF and 23 for GH.
$4x = 20$	Subtract 3 from each side.
$x = 5$	Divide each side by 4.

Exercise for Example 3

Find the value of the variable.

4. $AB = BC$, $BC = CD$

A 20 B C $7x - 1$ D

Copyright © McDougal Littell Inc.
All rights reserved.

NAME ____________________ DATE __________

Quick Catch-Up for Absent Students

For use with pages 88–94

The items checked below were covered in class on (date missed) __________

Lesson 2.6: Properties of Equality and Congruence (pp. 88–90)

____ **Goal:** Use properties of equality and congruence.

Material Covered:

____ Student Help: Look Back

____ Example 1: Name Properties of Equality and Congruence

____ Example 2: Use Properties of Equality

____ Student Help: Study Tip

____ Example 3: Justify the Congruent Supplements Theorem

____ Other (specify) ____________________

____

Homework and Additional Learning Support

____ Textbook exercises (teacher to specify) pp. 91–94 ____________________

____

____ Internet: More Examples at classzone.com

____ *Reteaching with Practice* worksheet

Copyright © McDougal Littell Inc.
All rights reserved.

NAME ______________________________ DATE ________

Quiz 2

For use after Lessons 2.4–2.6

In Exercises 1–3, use the diagram at right.

1. Find $m\angle 1$.
2. Find $m\angle 2$.
3. Find the value of y.

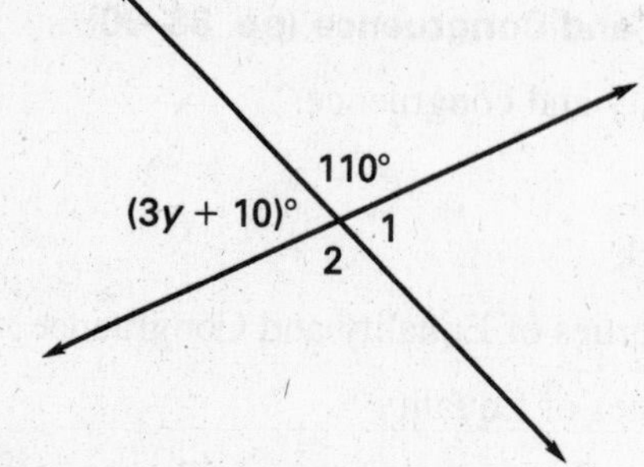

Answers

1. ________
2. ________
3. ________
4. See left.
5. See left.
6. See left.
7. ________
8. ________
9. ________
10. ________

In Exercises 4 and 5, rewrite the statement as an if-then statement. Then underline the hypothesis and circle the conclusion.

4. Vertical angles are congruent.

____

5. I will wear my sunglasses if it is sunny.

____

6. Write the if-then statement that follows from the pair of true statements.

   If it is summer, then the pool water is warm.

   If the pool water is warm, then Halie will go swimming.

____

Name the property that the statement illustrates.

7. If $m\angle X = m\angle Y$ and $m\angle Y = m\angle Z$, then $m\angle X = m\angle Z$.
8. If $\overline{MN} \cong \overline{KL}$, then $\overline{KL} \cong \overline{MN}$.
9. $\angle A \cong \angle A$
10. If $CD = DE$, then $\frac{CD}{8} = \frac{DE}{8}$.

Copyright © McDougal Littell Inc.
All rights reserved.

DATE ____________

...upport: Logic Puzzle

44

Wh...

2.

3.

4.

5.

6.

7. ...Alex has a guinea pig, then its name is Prince.

8. Fang is not the name of the dog.

	Alex	Carole	Mark	Sean	Tamara
Prince					
Stripe					
Dog					
Cat					
Fish					
Guinea Pig					
Rabbit					

After completing the logic puzzle, record your answers in the chart below.

Student's Name	Type of Pet	Pet's Name
Alex		
Carole		
Mark		
Sean		
Tamara		

Copyright © McDougal Littell Inc.
All rights reserved.

CHAPTER 2

NAME ______________________________ DATE ____________

Chapter Review Games and Activities

For use after Chapter 2

Complete the following crossword puzzle.

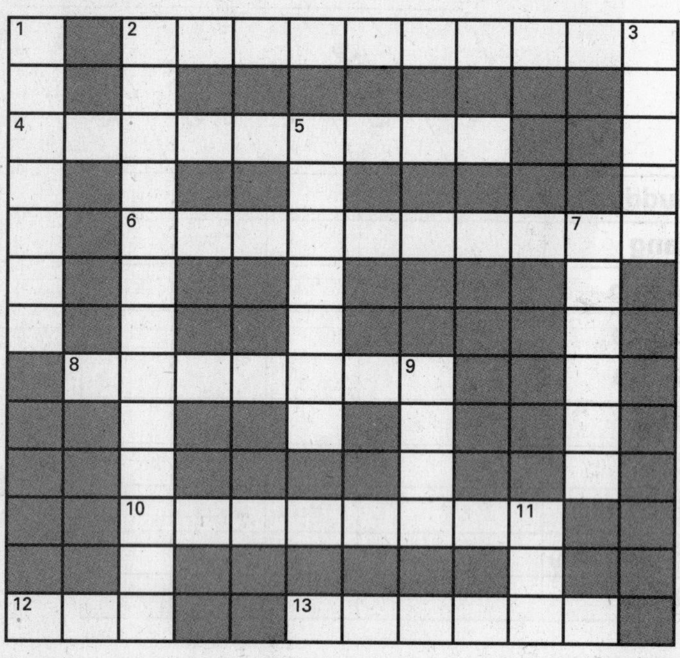

Across

2. Contains the "then" part of an if-then statement.
4. The property that states that if $a = b$, then $b = a$.
6. Pair of adjacent angles whose noncommon sides are on the same line.
8. A __?__ bisector is a segment, ray, line, or plane that intersects a segment at its midpoint.
10. Two angles that have a common vertex and side, but no common interior points.
12. An angle bisector is this kind of geometric figure.
13. Adjacent angles have a __?__ vertex.

Down

1. Divides into two congruent parts.
2. Two angles, the sum of whose measures is 90°.
3. Two angles are __?__ complementary and supplementary to the same angle.
5. The Congruent Supplements __?__ states that if two angles are supplementary to the same angle, then they are congruent.
7. This type of statement contains a hypothesis and a conclusion.
9. A theorem is a true statement that follows from other __?__ statements.
11. The midpoint of a segment divides it into this many congruent parts.

Copyright © McDougal Littell Inc.
All rights reserved.

CHAPTER 2

NAME ________________________________ DATE __________

Chapter Test A

For use after Chapter 2

M is the midpoint of the segment. Find the segment lengths.

1. Find AM and MC.

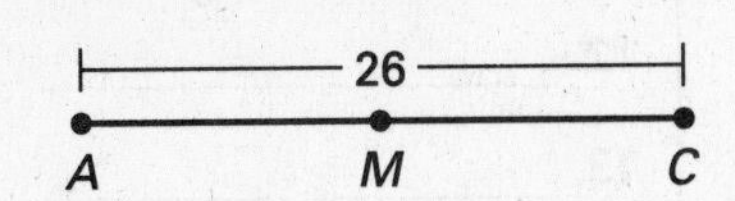

2. Find MT and RT.

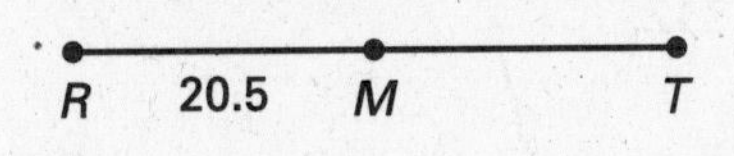

Find the coordinates of the midpoint of $\overline{KM}$.

3.

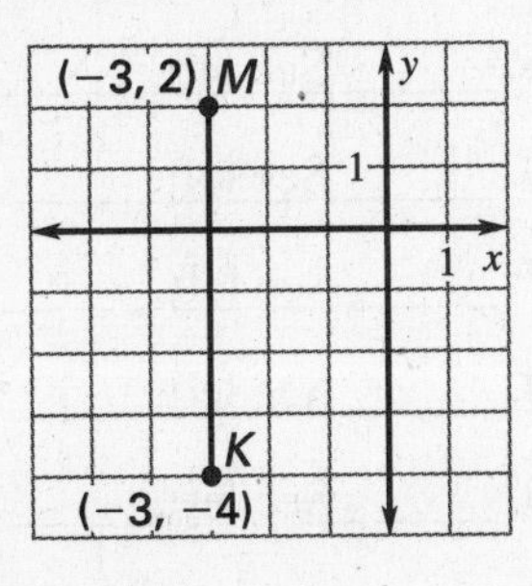

4.

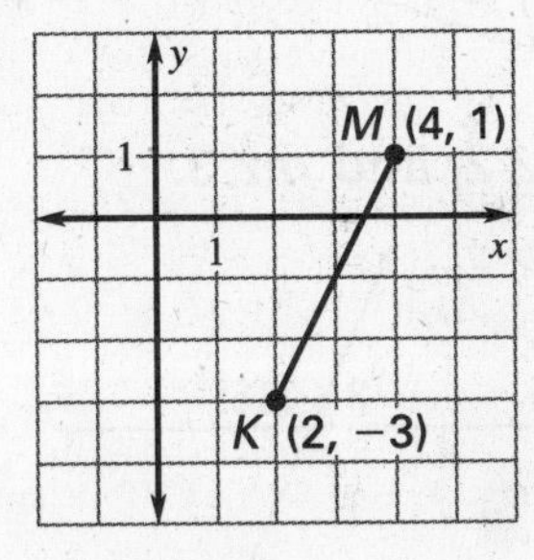

$\overrightarrow{ST}$ bisects $\angle RSU$. Find the angle measures.

5. Find $m\angle RST$ and $m\angle TSU$.

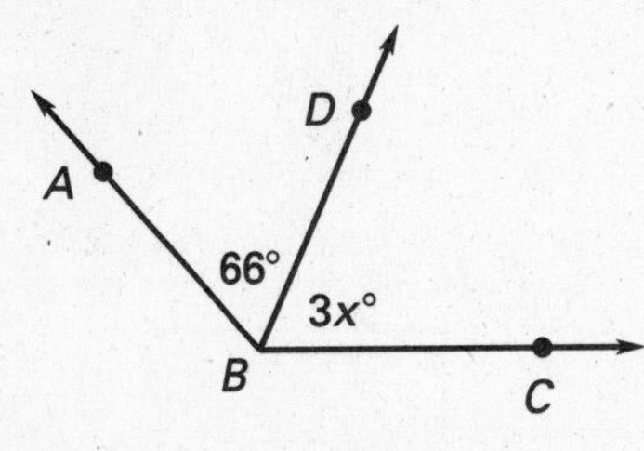

6. Find $m\angle TSU$ and $m\angle RSU$.

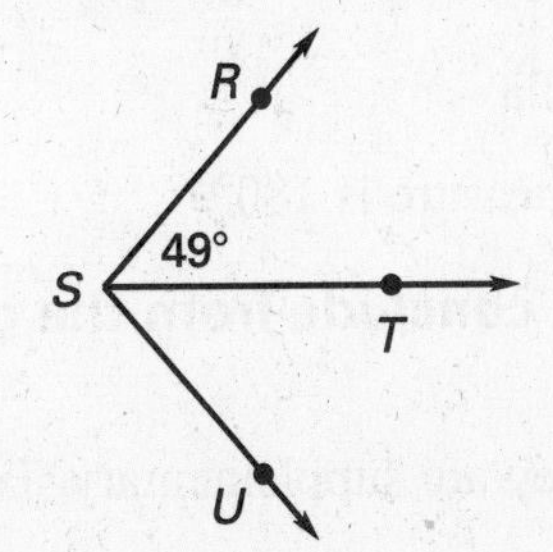

$\overrightarrow{BD}$ bisects $\angle ABC$. Find the value of the variable.

7.

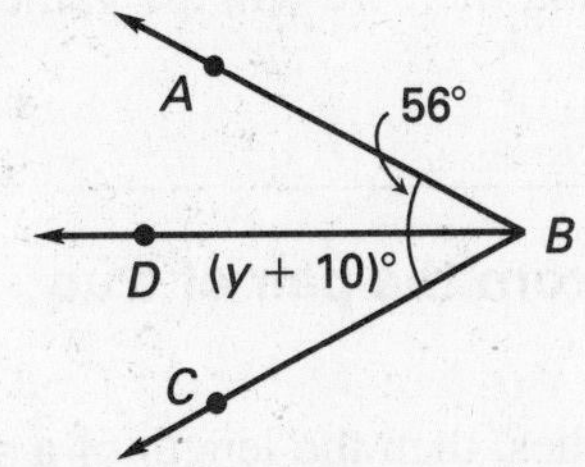

8.

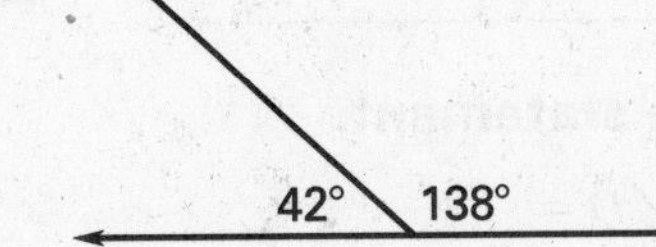

Determine whether the angles are *complementary, supplementary,* or *neither*. Then tell whether they are *adjacent* or *nonadjacent*.

9.

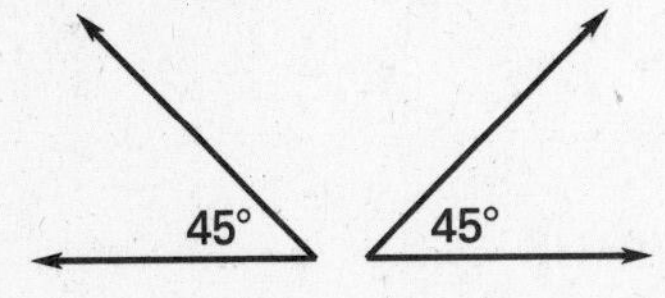

10.

Answers

1. ____________
2. ____________
3. ____________
4. ____________
5. ____________
6. ____________
7. ____________
8. ____________
9. ____________
10. ____________

Review and Assess

Copyright © McDougal Littell Inc.
All rights reserved.

NAME ______________________________ DATE __________

Chapter Test A

For use after Chapter 2

Find the measure of a complement and supplement of the angle.

11.

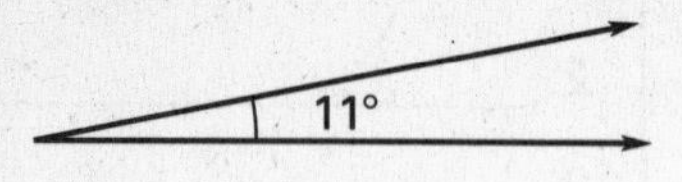

12.

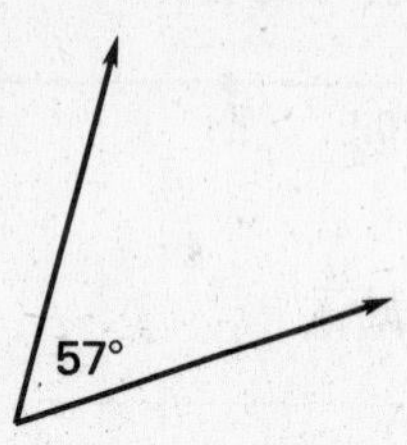

In Exercises 13 and 14, find $m\angle 1$, $m\angle 2$, and $m\angle 3$.

13.

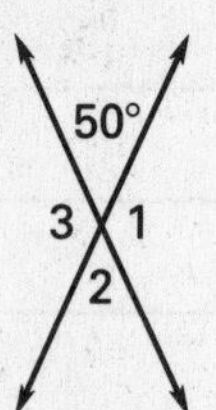

14.

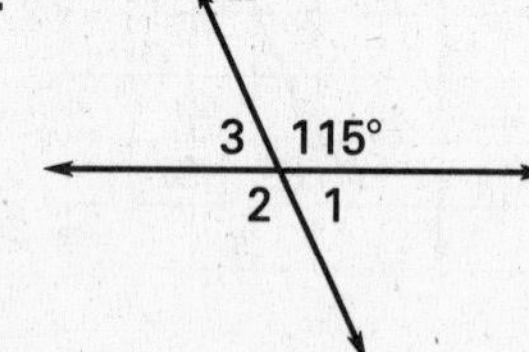

In Exercises 15 and 16, underline the hypothesis and circle the conclusion of the if-then statement.

15. If it rains, then you will need an umbrella.

16. If an angle is a straight angle, then its measure is 180°.

In Exercises 17 and 18, what can you conclude from the given true statements?

17. If two angles form a linear pair, then they are supplementary. Two angles form a linear pair.

Conclusion: ______________________________

18. If the basketball player makes the shot, then we win the game. The basketball player makes the shot.

Conclusion: ______________________________

Write the statement that follows from the pair of true statements.

19. If the perimeter of a square is 24 inches, then the length of a side is 6 inches.

If the length of a side of a square is 6 inches, then the area of the square is 36 square inches.

____

Use the property to complete the statement.

20. Reflexive Property of Equality: $m\angle D =$ ___?___

21. Transitive Property of Congruence:

If $\overline{PQ} \cong \overline{RS}$ and $\overline{RS} \cong \overline{TU}$, then ___?___ $\cong$ ___?___.

Answers

11. ______________

12. ______________

13. ______________

14. ______________

15. See left.

16. See left.

17. See left.

18. See left.

19. See left.

20. ______________

21. ______________

Copyright © McDougal Littell Inc.
All rights reserved.

NAME ______________________________ DATE __________

Chapter Test B

For use after Chapter 2

Line ℓ bisects the segment. Find the segment lengths.

1. Find KJ and KL.

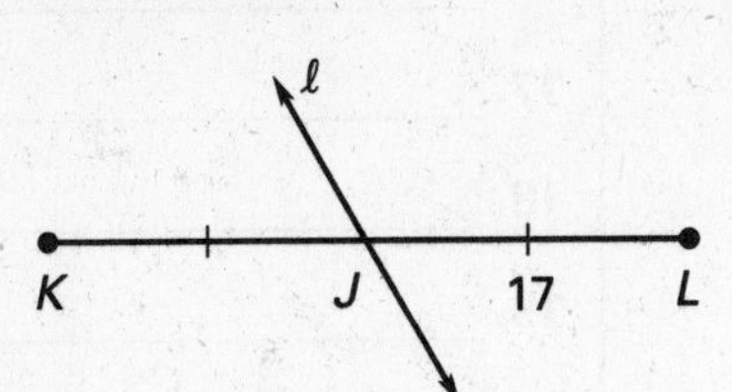

2. Find PR and PQ.

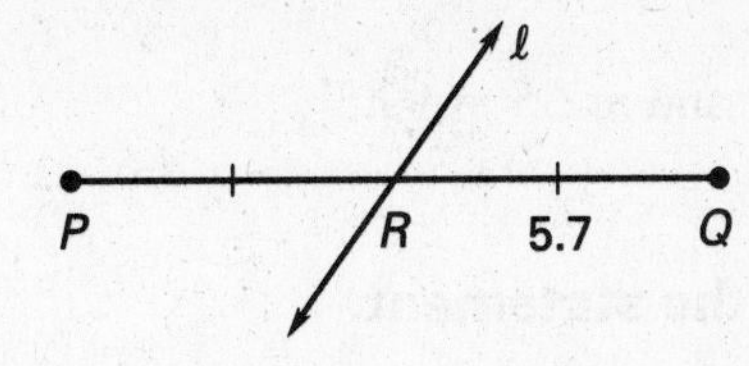

Find the coordinates of the midpoint of $\overline{AB}$.

3. $A(0, 0)$, $B(8, 2)$

4. $A(4, 7)$, $B(2, 13)$

5. $A(-3, -6)$, $B(5, 4)$

$\overrightarrow{QS}$ bisects $\angle PQR$. Find the angle measures.

6. Find $m\angle SQR$ and $m\angle PQR$.

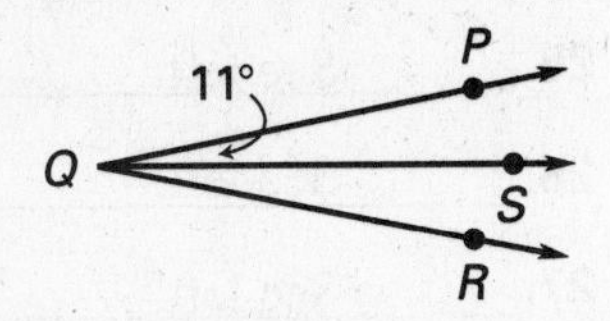

7. Find $m\angle PQS$ and $m\angle SQR$.

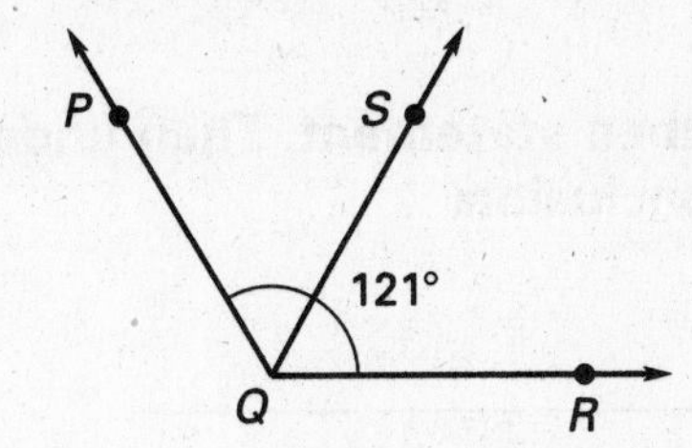

$\overrightarrow{EG}$ bisects $\angle DEF$. Find the value of x.

8.

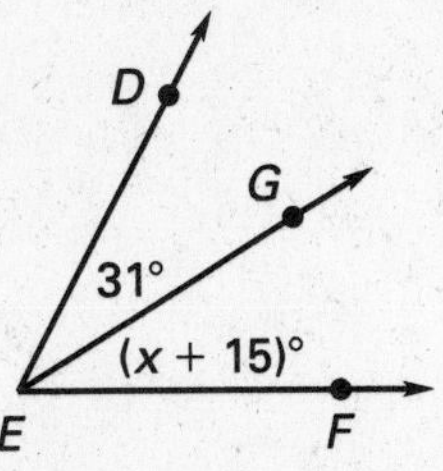

9.

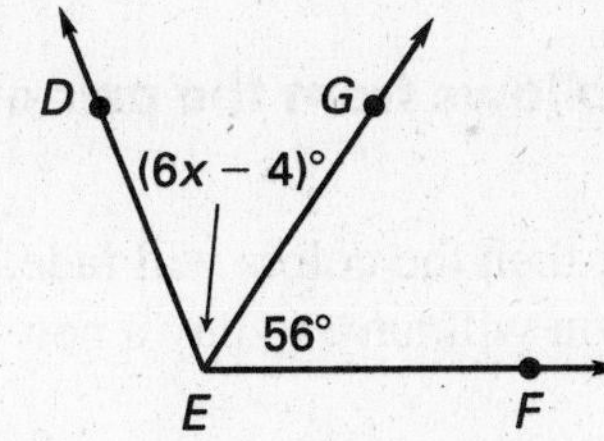

10. $\angle C$ is a complement of $\angle D$, and $m\angle C = 57°$. Find $m\angle D$.

11. $\angle E$ is a complement of $\angle F$, and $m\angle E = 9.5°$. Find $m\angle F$.

12. $\angle G$ is a supplement of $\angle H$, and $m\angle G = 37°$. Find $m\angle H$.

13. $\angle L$ is a supplement of $\angle M$, and $m\angle L = 114.5°$. Find $m\angle M$.

Find the value of the variable. Then use substitution to find $m\angle RSU$ and $m\angle UST$.

14.

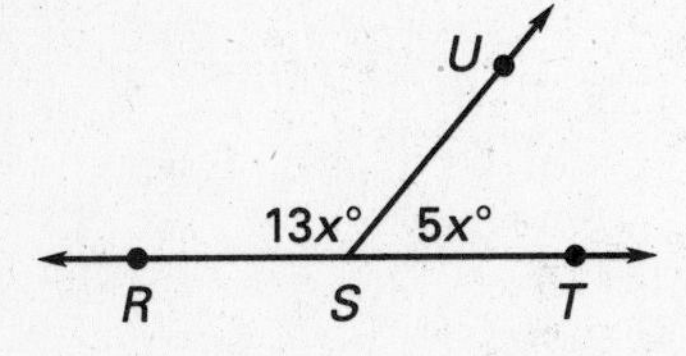

15.

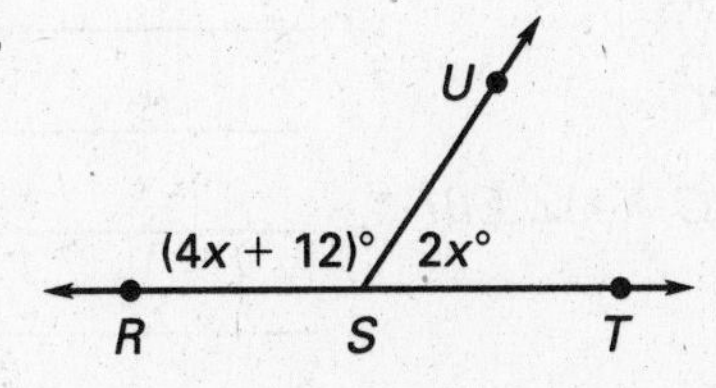

Answers

1. ____________
2. ____________
3. ____________
4. ____________
5. ____________
6. ____________
7. ____________
8. ____________
9. ____________
10. ____________
11. ____________
12. ____________
13. ____________
14. ____________

____
15. ____________

____

Copyright © McDougal Littell Inc.
All rights reserved.

NAME ______________________ DATE ________

Chapter Test B

For use after Chapter 2

Find the measure of the angle described.

16. $\angle 1$ and $\angle 2$ are a linear pair, and $m\angle 1 = 47°$. Find $m\angle 2$.

17. $\angle 5$ and $\angle 6$ are vertical angles, and $m\angle 5 = 99.5°$. Find $m\angle 6$.

Use the diagram to complete the statement.

18. $m\angle JHL =$ __?__ °

19. $m\angle JHN =$ __?__ °

20. $m\angle KHP =$ __?__ °

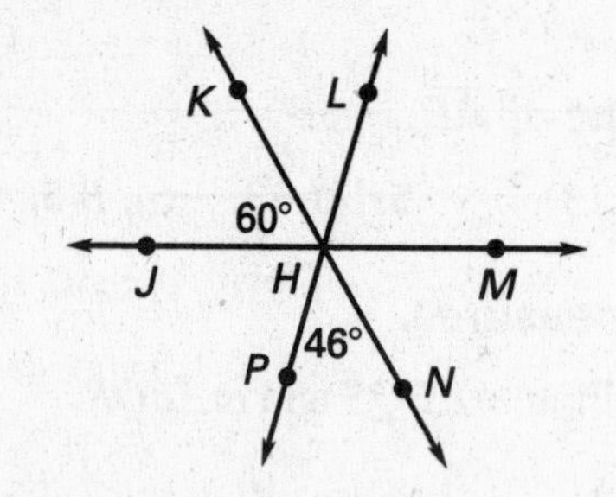

Rewrite the statement as an if-then statement. Then underline the hypothesis and circle the conclusion.

21. All ripe tomatoes are red.

____

22. You will need an umbrella if it is raining.

____

Write the if-then statement that follows from the pair of true statements.

23. If you wash your uniform in bleach, then the colors will fade. If the colors on your uniform fade, then you will have to buy a new one.

____

In the diagram, $m\angle ABE = m\angle FBC$. Complete the argument to show that $m\angle ABF = m\angle EBC$.

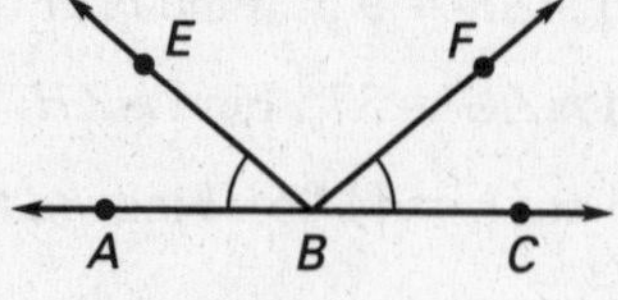

$m\angle ABE = m\angle FBC$	Given
24. $m\angle ABE + m\angle EBF = m\angle ABF$	________
25. $m\angle FBC + m\angle EBF = m\angle EBC$	________
26. $m\angle ABE + m\angle EBF = m\angle FBC + m\angle EBF$	________
27. $m\angle ABF = m\angle EBC$	________

Answers

16. ________
17. ________
18. ________
19. ________
20. ________
21. See left.
22. See left.
23. See left.
24. See left.
25. See left.
26. See left.
27. See left.

Copyright © McDougal Littell Inc.
All rights reserved.

CHAPTER 2

NAME ______________________________ DATE __________

SAT/ACT Chapter Test

For use after Chapter 2

1. M is the midpoint of $\overline{LN}$. Find the value of x.

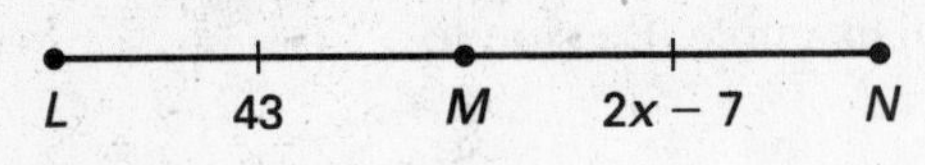

Ⓐ 19 Ⓑ 43

Ⓒ 25 Ⓓ 50

2. What are the coordinates of the midpoint of the segment joining (1, 1) and (−5, 7)?

Ⓐ (−4, 8) Ⓑ (−2, 4)

Ⓒ (−6, 6) Ⓓ (−3, 3)

3. $\overrightarrow{EF}$ bisects $\angle DEG$. What is $\angle DEF$?

Ⓐ 28.5°

Ⓑ 57°

Ⓒ 114°

Ⓓ 29.5°

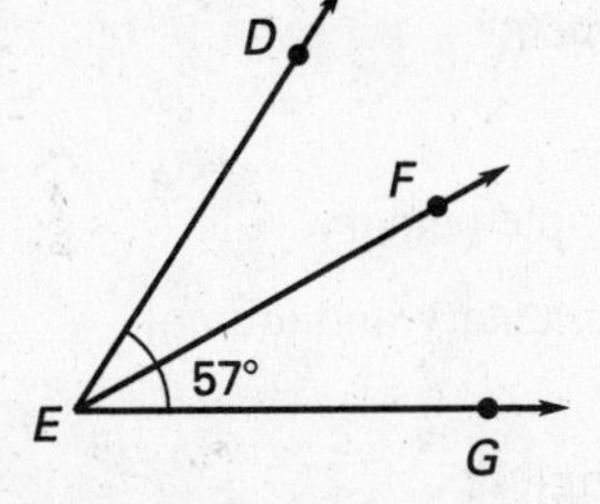

4. What is the measure of a supplement of a 41° angle?

Ⓐ 49° Ⓑ 139°

Ⓒ 59° Ⓓ 82°

5. $\angle 1$ and $\angle 2$ are both complementary to $\angle 3$, and $m\angle 2 = 73°$. What is $m\angle 1$?

Ⓐ 73° Ⓑ 17°

Ⓒ 34° Ⓓ 107°

6. What is $m\angle BEC$?

Ⓐ 142°

Ⓑ 76°

Ⓒ 112°

Ⓓ 114°

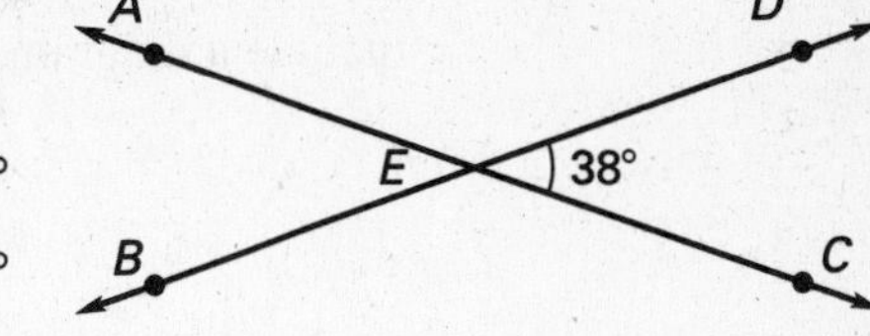

7. Use the diagram to find $m\angle FDG$.

Ⓐ 15°

Ⓑ 120°

Ⓒ 150°

Ⓓ 130°

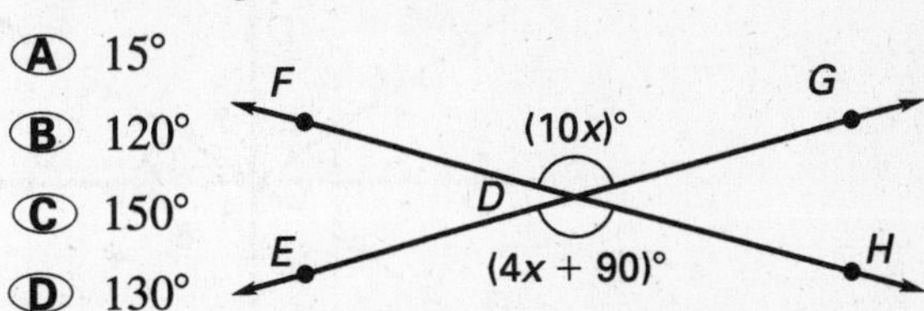

8. By the Law of Syllogism, what statement follows from the following pair of true statements?

If Nate eats chips, then he uses salsa dip.

If Nate uses salsa dip, then he gets indigestion.

Ⓐ Nate uses salsa dip.

Ⓑ Nate gets indigestion.

Ⓒ If Nate uses salsa dip, then he gets indigestion.

Ⓓ If Nate eats chips, then he gets indigestion.

9. Which statement illustrates the Transitive Property of Equality?

Ⓐ $LM = LM$

Ⓑ If $LM = NP$, then $NP = LM$.

Ⓒ If $LM = QR$, then $QR + LR = QR + LR$.

Ⓓ If $LM = NP$ and $NP = QR$, then $LM = QR$.

Review and Assess

Copyright © McDougal Littell Inc.
All rights reserved.

CHAPTER 2

NAME ______________________ DATE __________

Alternative Assessment and Math Journal

For use after Chapter 2

JOURNAL **1.** Write a letter to a friend explaining the Midpoint Formula. Explain the concept in words and provide a numerical example. Be sure to include a diagram.

MULTI-STEP PROBLEM **2.** Use the diagram below.

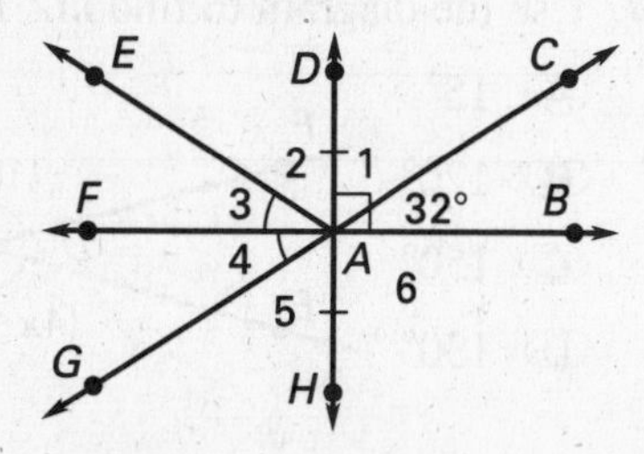

a. Point A is the midpoint of which segment?

b. Name an angle bisector of $\angle EAG$.

c. Name two pairs of angles that are complementary.

d. Name a pair of angles that are supplementary and are not congruent.

e. $\angle FAC$ and which angle are vertical angles?

f. Using the information given in the diagram, find the measure of each numbered angle.

g. ***Logical Reasoning*** You know that $\angle 2$ and $\angle 3$ are complements, $\angle 4$ and $\angle 5$ are complements, and $m\angle 3 = m\angle 4$. Complete the argument below to show that $m\angle 2 = m\angle 5$.

$\angle 2$ and $\angle 3$ are complements. $\angle 4$ and $\angle 5$ are complements.	Given
$m\angle 2 + m\angle 3 = 90°$ $m\angle 4 + m\angle 5 = 90°$	Definition of _?_
$m\angle 2 + m\angle 3 = m\angle 4 + m\angle 5$	Transitive Property of Equality
$m\angle 3 = m\angle 4$	_?_
$m\angle 2 + m\angle 4 = m\angle 4 + m\angle 5$	_?_ Property of Equality
$m\angle 2 + m\angle 5$	_?_ Property of Equality

Review and Assess

Copyright © McDougal Littell Inc. All rights reserved.

NAME ______________________ DATE __________

Alternative Assessment Rubric

For use after Chapter 2

JOURNAL SOLUTION

1. Complete answers should include:
   - A clear and concise explanation of the Midpoint Formula.
   - An example, done correctly, using the Midpoint Formula.
   - A diagram clearly showing the endpoints and midpoint of a segment.

MULTI-STEP PROBLEM SOLUTION

2. **a.** $\overline{DH}$

   b. $\overrightarrow{AF}$

   c. *Sample answer:* $\angle DAC$ and $\angle CAB$, $\angle DAE$ and $\angle EAF$

   d. *Sample answer:* $\angle FAC$ and $\angle CAB$

   e. $\angle GAB$

   f. $m\angle 1 = 58°$, $m\angle 2 = 58°$, $m\angle 3 = 32°$, $m\angle 4 = 32°$, $m\angle 5 = 58°$, $m\angle 6 = 90°$

   g. complement; Given; Substitution; Subtraction

MULTI-STEP PROBLEM RUBRIC

4 Students answer all parts of the problem correctly and completely, showing all work.

3 Students answer all parts of the problem. Students may have a small mathematical error. Part (g) may contain an incorrect reason.

2 Students answer all parts of the problem. Students may have several mathematical errors. Reasons in part (g) may be incorrect.

1 Students do not complete the problem. Solutions are incorrect.

Copyright © McDougal Littell Inc.
All rights reserved.

CHAPTER 2

NAME ______________________________ DATE ____________

Project: Origami

For use after Chapter 2

OBJECTIVE **Create an origami figure and analyze the angles formed in the figure.**

MATERIALS square pieces of paper, reference books

INVESTIGATION Origami is the ancient Japanese art of paper folding. In origami, one piece of paper is folded in different ways to form objects like birds, butterflies, flowers, planes, and boxes. Although some of the finished pieces of origami look simple, the instructions for folding the origami can be complicated. The instructions used to create an origami figure are called *diagrams*. An example of an origami diagram and the completed figure are shown below.

A Simple Airplane

1. Fold and unfold.

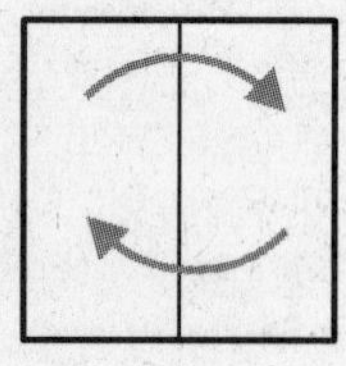

2. Upper right corner meets the end of the crease.

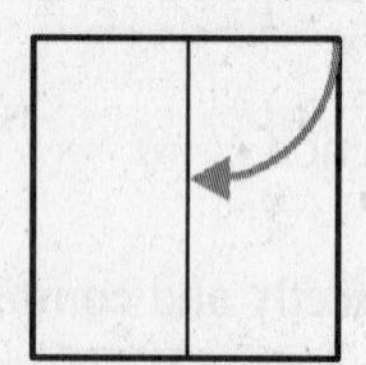

3. Upper left corner meets the end of the crease.

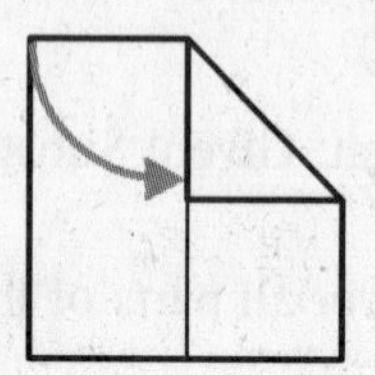

4. Fold upper right side to meet center.

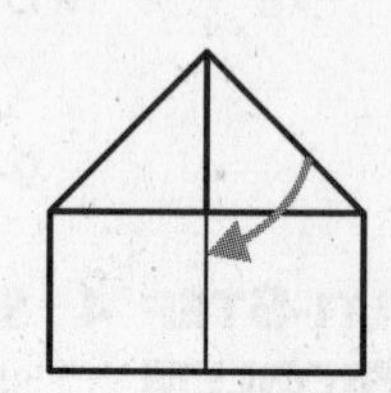

5. Fold upper left side to meet center.

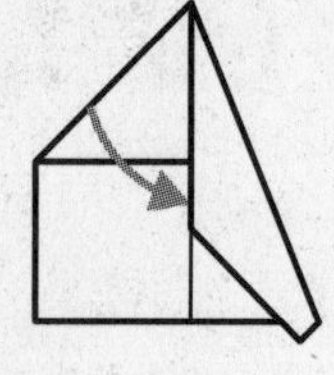

6. Fold in half.

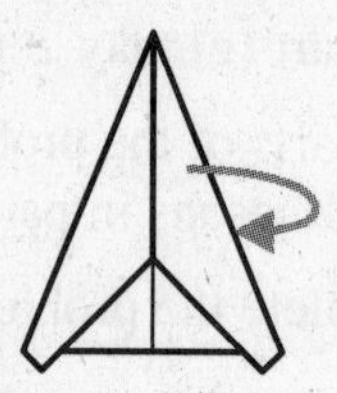

7. Place fingers under flaps.

For this project you will be finding and creating your own origami figure from a diagram. When you find the origami figure you want to create, make sure that you understand all of the directions. You may need to look in a reference book or website to figure out how to make certain folds.

1. Find an origami figure that you like. Origami figures can be quite complex, so you should start with something that has a diagram with a small number of steps and simple folds. You can do this by going to the library and finding a book of origami or by searching for origami figures on the Internet.
2. Copy or print out the diagram for your figure. Then construct the origami figure described by the diagram.
3. Look at your origami diagram again. Find and label a step that shows an example of a bisected angle being formed from a fold. Then find and label a step that shows an example of a pair of complementary angles and a pair of supplementary angles being formed from a fold.

PRESENT YOUR RESULTS Your project report should include your origami diagram, the source for your origami diagram, your results to Exercise 3, and the origami figure. In your own words, describe the different types of folds, if specified, that you needed to learn about in order to construct your origami figure.

Copyright © McDougal Littell Inc.
All rights reserved.

CHAPTER 2 CONTINUED

NAME ______________________ DATE __________

Project: Origami

For use after Chapter 2

GOALS
- Identify bisected angles.
- Identify pairs of complementary and supplementary angles.

MANAGING THE PROJECT Students will need to visit the library or have access to the Internet. Resource books could be made available in the classroom.

Guiding Students' Work This project will probably require that students work on it during nonclass time. It will take some time for them to choose an origami figure to construct. You may want to check on their choices to make sure that they are trying to create a relatively simple figure. Make sure that before they try folding their figure, they read the diagram and define any folds required in the directions. Some common folds are called valley folds, mountain folds, and reverse folds. Each student should have several pieces of paper, as it may take a couple of tries to get a neatly folded and accurately constructed final figure.

Concluding the Project Consider a classroom display of the figures constructed by the students. You could also have them find new diagrams, or exchange diagrams with each other to create more figures.

RUBRIC The following rubric can be used to assess student work.

4 The student includes a neatly constructed origami figure, the origami diagram, and the source of the origami design. The student correctly identifies a bisected angle, a pair of complementary angles, and a pair of supplementary angles. The descriptions of the different types of folds used are clear and detailed.

3 The student includes the diagram and its source, but the origami figure is a bit sloppy in construction. The student correctly identifies a bisected angle, a pair of complementary angles, and a pair of supplementary angles. The descriptions of the different types of folds are sufficient, but could be better detailed.

2 The student includes the diagram and its source, but the origami figure is sloppy in construction. The student does not correctly identify either a bisected angle, a pair of complementary angles, or a pair of supplementary angles. The descriptions of the different types of folds are included, but some of the details are missing or confusing.

1 The student includes the diagram, but the source is missing and/or the origami figure is poorly constructed. The student does not correctly identify a bisected angle, a pair of complementary angles, and/or a pair of supplementary angles. The descriptions of the different types of folds are missing.

Copyright © McDougal Littell Inc.
All rights reserved.

CHAPTER 2

NAME ______________________ DATE __________

Cumulative Review

For use after Chapters 1–2

Use the diagram at the right. (Lesson 1.3)

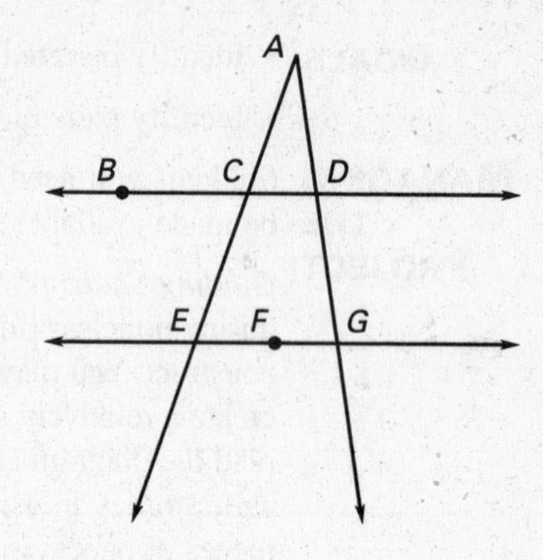

1. Name a point that is collinear with D and G.
2. Name two lines that pass through E.
3. Name the points that are not collinear with D and C.

Sketch the figure described. (Lessons 1.3 and 1.4)

4. Three noncollinear points, A, B, and C. Draw $\overleftrightarrow{AB}$ and $\overleftrightarrow{AC}$.
5. Two planes that do not intersect, and line r that intersects both.

Find the length. (Lesson 1.5)

6. Find MO.

7. Find PQ.

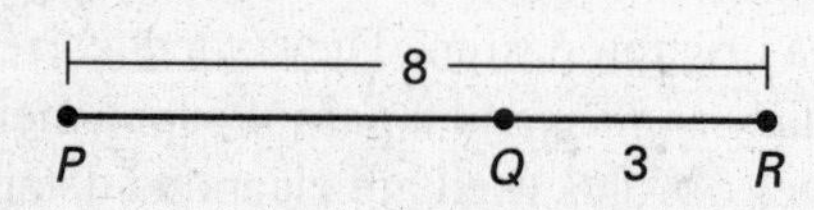

Classify the angle as *acute, right, obtuse* or *straight*. (Lesson 1.6)

8. $m\angle A = 180°$ 9. $m\angle B = 97°$ 10. $m\angle C = 32°$ 11. $m\angle D = 90°$

Find the coordinates of the midpoint of $\overline{AB}$. (Lesson 2.1)

12. $A(12, -4), B(8, -2)$ 13. $A(-7, 5), B(1, -3)$

$\overrightarrow{LN}$ bisects $\angle KLM$. Find the angle measure. (Lesson 2.2)

14. Find $m\angle KLN$.

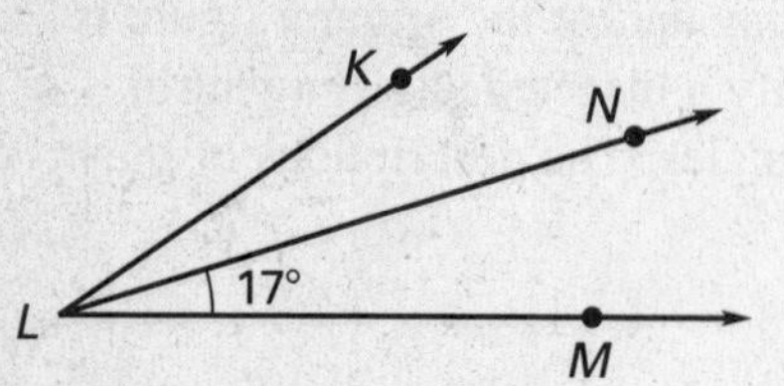

15. Find $m\angle NLM$.

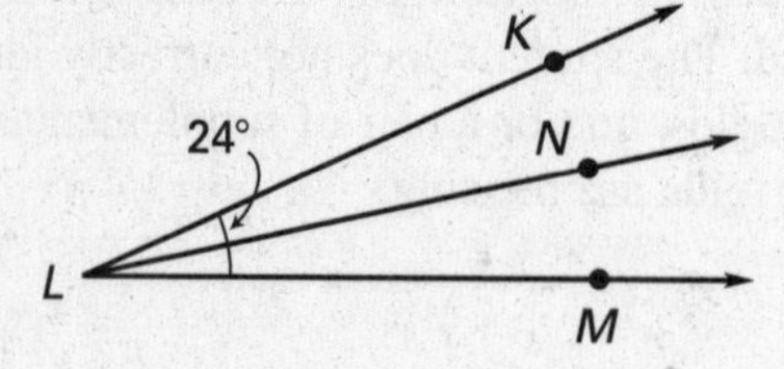

16. Find $m\angle KLM$.

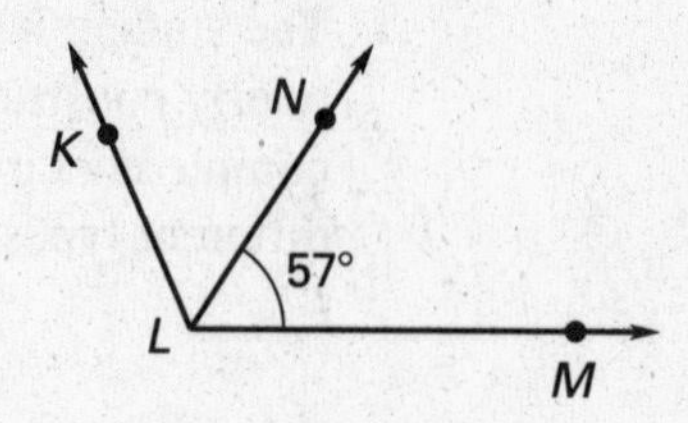

Review and Assess

Copyright © McDougal Littell Inc.
All rights reserved.

CHAPTER 2 CONTINUED

NAME ____________________ DATE __________

Cumulative Review

For use after Chapters 1–2

$\overrightarrow{QS}$ bisects $\angle PQR$. Find the value of the variable. (Lesson 2.2)

17.

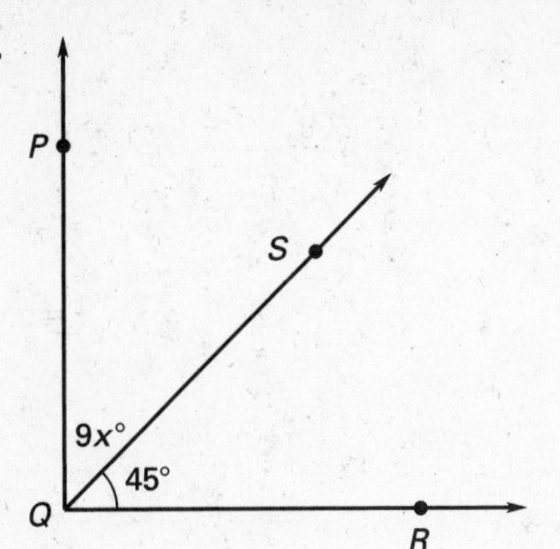

18.

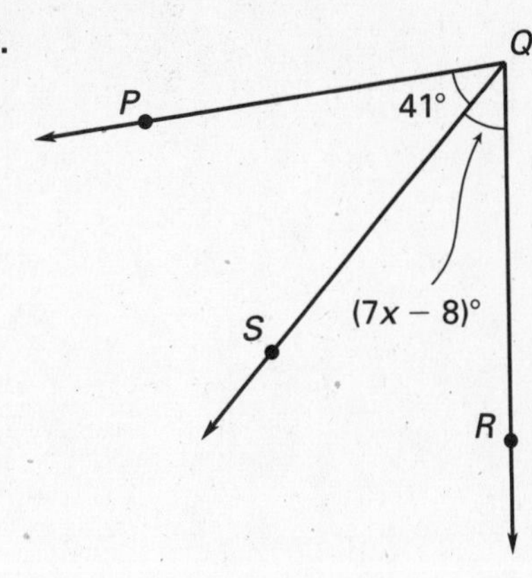

19. $\angle A$ is a supplement of $\angle B$, and $m\angle A = 52°$. Find $m\angle B$. *(Lesson 2.3)*

20. $\angle X$ is a complement of $\angle Y$, and $m\angle Y = 37°$. Find $m\angle X$. *(Lesson 2.3)*

Use the diagram at the right. (Lesson 2.4)

21. Find $m\angle 1$.

22. Find $m\angle 2$.

23. Find $m\angle 3$.

24. Name two pairs of vertical angles.

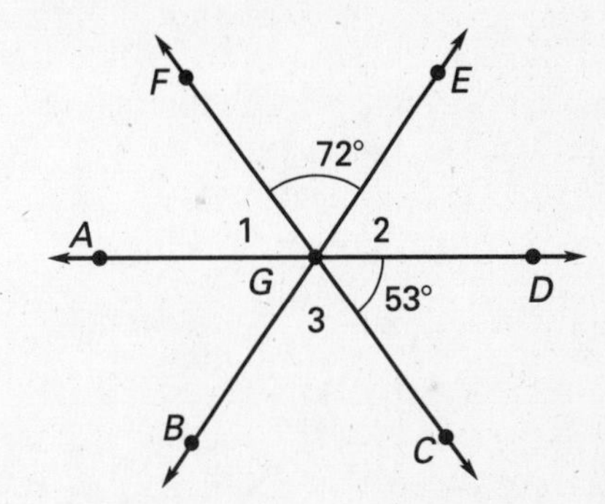

25. Find the value of the variable. *(Lesson 2.4)*

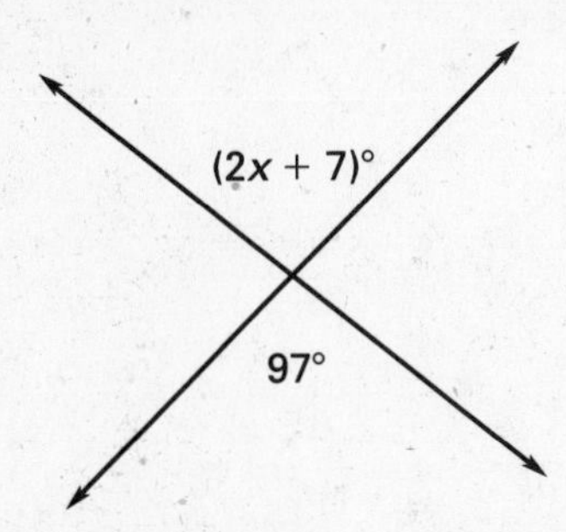

26. What can you conclude from the given true statements? *(Lesson 2.5)*

If you leave windows open when it is raining, then your floors will get wet.

You leave your windows open when it is raining.

Name the property the statement illustrates. (Lesson 2.6)

27. If $m\angle A = m\angle X$, then $m\angle X = m\angle A$.

28. $\angle MNO \cong \angle MNO$

29. If $AB = CD$ and $CD = FG$, then $AB = FG$.

30. If $x = 10$ then $x + y = 10 + y$.

Review and Assess

Copyright © McDougal Littell Inc.
All rights reserved.

ANSWERS

Answers

Chapter Support

Parent Guide

2.1: (2, 2) **2.2:** 55° **2.3:** 54°, 144° **2.4:** 135°, 45°, 135° **2.5:** If Roger works, then he buys new clothes. **2.6:** $m\angle K = m\angle M$

Strategies for Reading Mathematics

1. $\angle 2$ and $\angle 3$, $\angle 3$ and $\angle 4$, $\angle 4$ and $\angle 1$; each pair shares a common vertex and side but has no common interior points. **2.** $\angle 2$ and $\angle 3$, $\angle 3$ and $\angle 4$, $\angle 4$ and $\angle 1$; the two angles in each pair are adjacent with noncommon sides that are opposite rays. **3.** $\angle 2$ and $\angle 4$; their sides form two pairs of opposite rays. **4.** $\angle 5$ and $\angle 7$, $\angle 6$ and $\angle 8$; *Sample answer*: look at the relative position of the vertical angles in the drawing in the notebook and find angles that are in the same relative position.

Lesson 2.1

Warm-Up Exercises

1. $z = 7$ **2.** $x = 11$ **3.** $y = 22$ **4.** $(2, -1)$ **5.** $(-1, 2)$

Daily Homework Quiz

1. 30° **2.** 90° **3.** $\angle ABC$ is acute, $\angle CBD$ is acute, and $\angle ABD$ is right.

Practice A

1. midpoint **2.** segment bisector **3.** bisect **4.** $TM = 7$; $MR = 7$ **5.** $FM = 18$; $MD = 18$ **6.** $MR = 5$; $QR = 10$ **7.** $KM = 10$; $KL = 20$ **8.** (4, 2) **9.** (−3, 3)

10.

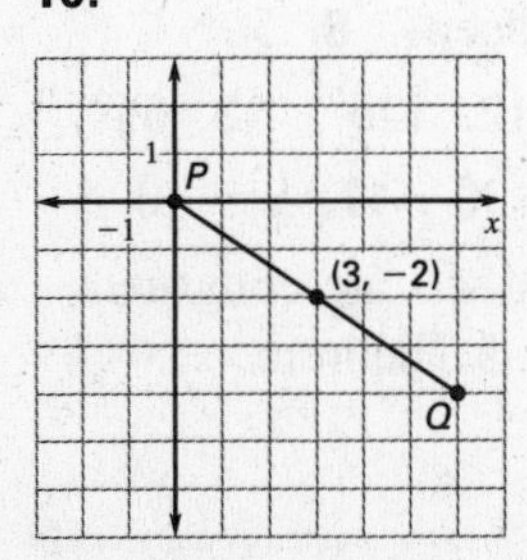

11.

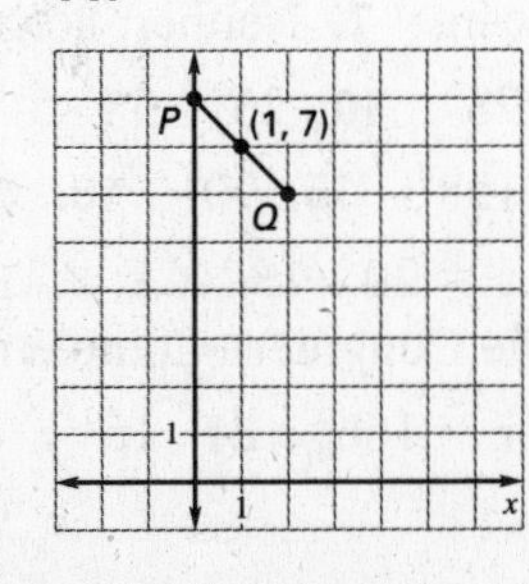

12.

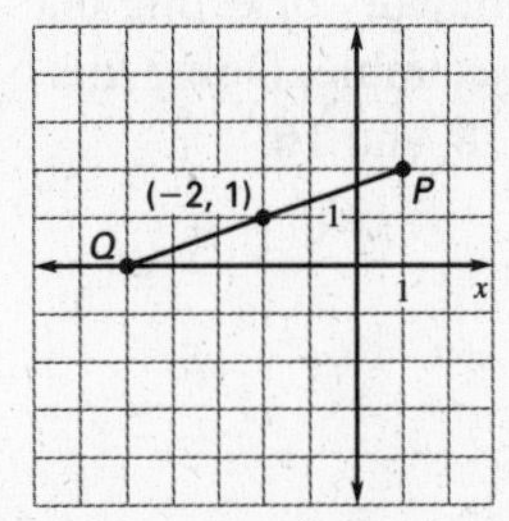

13. $x = 9$ **14.** $x = 8$ **15.** 250 cm

Practice B

1. M is the midpoint because it divides $\overline{JK}$ into two congruent segments. **2.** M is not the midpoint because M does not lie on $\overline{JK}$. **3.** M is not the midpoint because it is an endpoint of the segment. **4.** $PM = 11$; $MQ = 11$ **5.** $TM = 7.5$; $MS = 7.5$ **6.** $AN = 9.2$; $AB = 18.4$ **7.** $FG = 17$; $EG = 34$ **8.** $x = 3$ **9.** $x = 5$ **10.** (1, 1) **11.** $\left(\frac{5}{2}, -4\right)$ **12.** $(-4, 4)$ **13.** $\left(\frac{7}{2}, -3\right)$ **14.** $\left(2, \frac{5}{2}\right)$ **15.** $(-2, 4)$ **16.** $5\frac{1}{2}$ feet

Reteaching with Practice

1. $x = 7$ **2.** $x = 7$ **3.** $x = 2$ **4.** $x = 7.5$ **5.** $AM = 5$, $MB = 5$ **6.** $ML = 21$, $KL = 42$ **7.** $ST = 14$, $SM = 7$ **8.** $CM = 12.5$, $MD = 12.5$

9. $M(3, 5)$

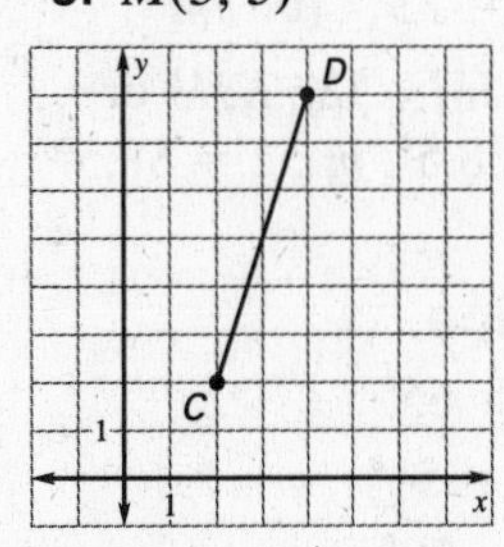

10. $M(-2, 1)$

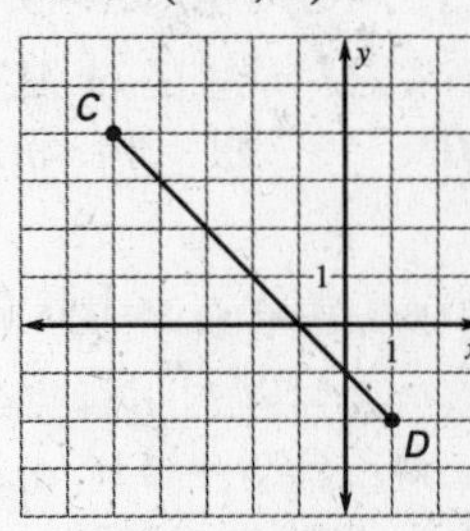

11. $M(2.5, -5)$

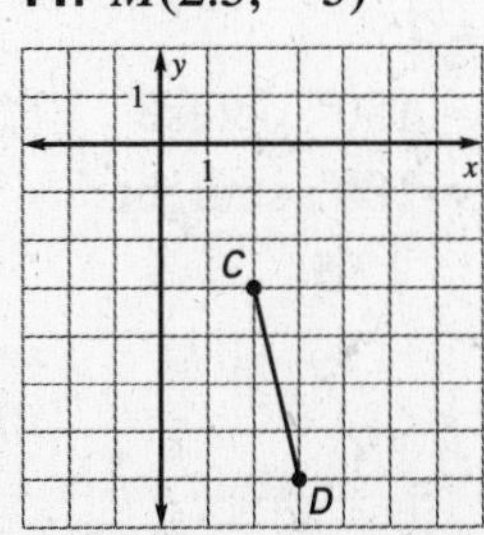

Copyright © McDougal Littell Inc.
All rights reserved.

Lesson 2.1 *continued*

Learning Activity

1. 4 inches; 4 inches **2.** $AB = 2AC$

3. *Sample answer:* Continually bisecting one inch is convenient and measurable. Doing this creates increments of 2, 4, 8, and 16.

Lesson 2.2

Warm-Up Exercises

1. $m\angle A = 116°$ **2.** $m\angle B = 65°$

3. $m\angle C = 52°$ **4.** $x = 9$ **5.** $x = 15$

Daily Homework Quiz

1. $GH = 27.5$, $FH = 55$ **2.** $x = 5$ **3.** $M(0, 2)$

4. 1.7 miles

Practice A

1. $\angle ABC$; $\overrightarrow{BD}$ **2.** twice **3.** 23° **4.** 70°

5. 45° **6.** $m\angle CBA = 35°$; $m\angle DBC = 70°$

7. $m\angle CBA = 75°$; $m\angle DBC = 150°$

8. $m\angle CBA = 45°$; $m\angle DBC = 90°$ **9.** $x = 30$

10. $x = 20$ **11.** $x = 3$ **12.** false **13.** true

14. true **15.** false **16.** 36°

Practice B

1. 65° **2.** 10° **3.** 74.5°

4. $m\angle EFJ = 45°$; $m\angle JFG = 45°$

5. $m\angle EFJ = 67.5°$; $m\angle JFG = 67.5°$

6. $m\angle EFJ = 16°$; $m\angle JFG = 16°$

7. $x = 25$ **8.** $x = 5$ **9.** $x = 23$ **10.** $\overrightarrow{JM}$

11. $KJM = MJL$ or $MJK = LJM$ **12.** KJM or MJK **13.** 92 **14.** $m\angle 1 = 58.5°$; $m\angle 2 = 58.5°$

Reteaching with Practice

1. $m\angle APS = 65°$, $m\angle SPB = 65°$

2. $m\angle CPS = 51°$, $m\angle DPS = 51°$

3. $m\angle EPS = 90°$, $m\angle SPF = 90°$

4. $m\angle GPS = 75°$, $m\angle HPS = 75°$

5. $m\angle CBD = 75°$, $m\angle ABC = 150°$; obtuse

6. $m\angle CBD = 90°$, $m\angle ABC = 180°$; straight

7. $x = 55$ **8.** $x = 25$ **9.** $x = 12.5$

Real-Life Application

1.

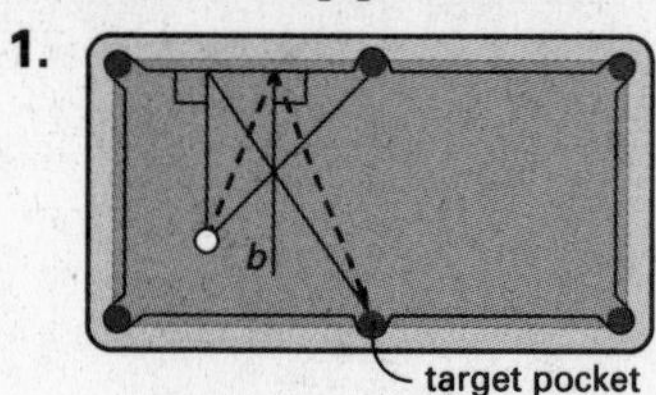

2.

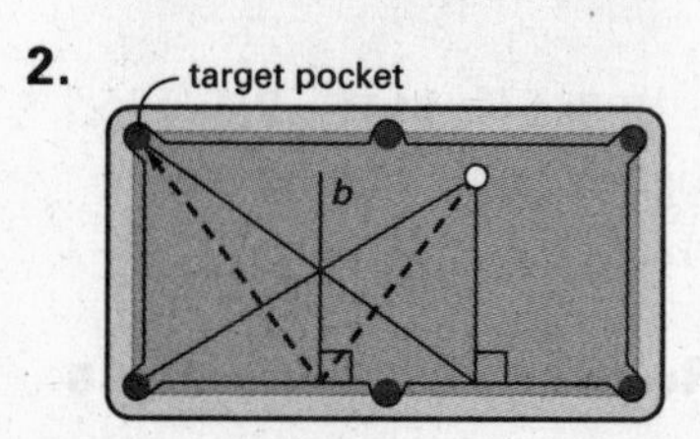

Lesson 2.3

Warm-Up Exercises

1. $\angle DBE$ or $\angle EBC$ **2.** $\angle DBA$ or $\angle DBC$

3. $\angle ABE$ **4.** $\angle ABC$

Daily Homework Quiz

1. $m\angle ABD = m\angle DBC = 77°$

2. $m\angle GHL = 64°$, $m\angle GHJ = 128°$; obtuse

3. $x = 7$ **4.** $y = 10$

Practice A

1. False; two angles are complementary if the sum of their measures is 90°. Or, two angles are supplementary if the sum of their measures is 180°. **2.** true **3.** False; two angles are adjacent angles if they share a common vertex and a common side, but have no common interior points.

4. true **5.** complementary

6. supplementary **7.** neither **8.** 58° **9.** 75°

10. 13° **11.** 20° **12.** 90° **13.** 162° **14.** $\angle 2$

15. $\angle 2$ **16.** $\angle 1$, $\angle 3$ **17.** 60°

Practice B

1. 90 **2.** 180 **3.** adjacent **4.** theorem

5. Supplementary; nonadjacent **6.** Neither; adjacent **7.** Neither; nonadjacent **8.** 50°

9. 78° **10.** 26° **11.** 5° **12.** 126° **13.** 65°

14. 158° **15.** 90° **16.** $x = 30$ **17.** $x = 10$

18. $x = 20$ **19.** $\angle 1$; $\angle 1$ and $\angle 3$ are congruent by the Congruent Complements Theorem.

20. $x = 106°$ **21.** 16°

Copyright © McDougal Littell Inc.
All rights reserved.

Lesson 2.3 *continued*

Reteaching with Practice

1. complementary **2.** supplementary **3.** neither **4.** adjacent **5.** adjacent **6.** nonadjacent **7.** $m\angle B = 79°$ **8.** $m\angle A = 58°$ **9.** $m\angle D = 2°$

Real-Life Application

1. 90°; the angles form the corner of a rectangle, which is a right angle. **2.** No, no; If the width of the top frame is different from the width of the side frames, then the angle pairs are not congruent. However, if the top and sides are the same width, then $\angle 1$, $\angle 2$, $\angle 3$, and $\angle 4$ must all have measures of 45°. **3.** 40° **4.** 52°, 38°, 38°

5. Overlapping at the corners and then cutting is an easy way to get complementary angles that match.

Quiz 1

1. 15; 15 **2.** 7; 14 **3.** 9 **4.** $(6, -1)$ **5.** $(3, 2)$ **6.** 22° **7.** 54.5° **8.** $x = 9$ **9.** complementary **10.** neither **11.** neither **12.** 34°; 124°

Lesson 2.4

Warm-Up Exercises

1. nonadjacent **2.** adjacent **3.** 113° **4.** 32°

Daily Homework Quiz

1. complementary; adjacent **2.** neither; nonadjacent **3.** 63°; 153° **4.** 79°; 169° **5.** $x = 7$

Practice A

1. supplementary **2.** congruent **3.** vertical angles **4.** neither **5.** linear pair **6.** 135° **7.** 95° **8.** 150° **9.** $\angle 3 \cong \angle 1$ **10.** $\angle 4 \cong \angle 1$ **11.** $\angle 2 \cong \angle 1$ **12.** $m\angle 1 = 30°$; $m\angle 2 = 150°$; $m\angle 3 = 30°$ **13.** $m\angle 1 = 126°$; $m\angle 2 = 54°$; $m\angle 3 = 126°$ **14.** $m\angle 1 = 135°$; $m\angle 2 = 45°$; $m\angle 3 = 135°$ **15.** $x = 16$ **16.** $x = 63$ **17.** $x = 10$ **18.** $\angle 1$ and $\angle 3$; $\angle 2$ and $\angle 4$ **19.** 110° **20.** 70° **21.** 110°

Practice B

1. $\angle 2$; $\angle 4$ *or* $\angle 4$; $\angle 2$ **2.** $\angle 4$ **3.** 30° **4.** $\angle 2$ **5.** 29° **6.** 154° **7.** 89° **8.** $m\angle 1 = 145°$; $m\angle 2 = 35°$; $m\angle 3 = 145°$ **9.** $m\angle 1 = 90°$; $m\angle 2 = 90°$; $m\angle 3 = 90°$ **10.** $m\angle 1 = 52°$; $m\angle 2 = 128°$; $m\angle 3 = 52°$ **11.** 96 **12.** 84 **13.** 28 **14.** 124 **15.** $x = 10$; $m\angle PQR = 39°$ **16.** $y = 34$; $m\angle PQR = 104°$ **17.** $x = 12$; $m\angle PQR = 120°$ **18.** $m\angle 2 = 63.4°$; $m\angle 3 = 63.4°$; $m\angle 4 = 26.6°$; $m\angle 5 = 26.6°$; $m\angle 6 = 63.4°$; $m\angle 7 = 63.4°$; $m\angle 8 = 26.6°$

Reteaching with Practice

1. neither **2.** vertical angles **3.** linear pair **4.** $x = 123$ **5.** $x = 88$ **6.** $x = 55$ **7.** $x = 126$ **8.** $x = 90$ **9.** $y = 9$ **10.** $z = 11$

Lesson 2.5

Warm-Up Exercises

1. true **2.** false **3.** false

Daily Homework Quiz

1. $\angle 1$ and $\angle 5$ or $\angle 4$ and $\angle 5$ **2.** $\angle 1$ and $\angle 4$ **3.** $m\angle 3 = 100°$, $m\angle 4 = 35°$, and $m\angle 5 = 145°$ **4.** $x = 7$ **5.** $y = 9$

Practice A

1. Two angles have the same measure; the angles are congruent. **2.** Two angles form a linear pair; the angles are supplementary. **3.** The sum of the measures of two angles is 90°; the angles are complementary. **4.** The measure of an angle is 90°; the angle is a right angle. **5.** If a school yearbook costs less than $20, then I will purchase one. **6.** If your team wins the district championship, then your team will travel to the state championship game. **7.** If your bicycle has a flat tire, then you cannot ride your bicycle.

8. If it snows six inches overnight, then school will be cancelled. **9.** $3x + 1$ has a value of 13.

Copyright © McDougal Littell Inc.
All rights reserved.

Lesson 2.5 *continued*

10. An angle that has a measure of 51° is an acute angle. **11.** If a number is divisible by 4, then the number is even. **12.** If the perimeter of a square is 12 centimeters, then the area of the square is 9 square centimeters.

13. If it rains, then the food will go to waste.

14. If you want to look younger, then try our Creme de Youth for thirty days. **15.** If you want a better job, then enroll in Seville College's on-line degree program.

Practice B

1. Two planes intersect; their intersection is a line. **2.** Angle A is acute; the measure of angle A is between 0° and 90°. **3.** The sum of the measures of two angles is 180°; the angles are supplementary. **4.** The measure of an angle is between 90° and 180°; the angle is obtuse.

5. If two angles have the same measure, then the angles are congruent. **6.** If two angles form a linear pair, then the angles are supplementary.

7. If an angle has a measure of 90°, then the angle is a right angle. **8.** If an angle has a measure between 90° and 180°, then the angle is an obtuse angle. **9.** If a dog has proper training, then the dog will not misbehave. **10.** Law of Detachment; you conclude that you can buy a new CD. **11.** Law of Syllogism; you conclude if the area of a square is 49 in.^{2}, then the perimeter of the square is 28 inches. **12.** The area of the rectangle is 10 cm^2. **13.** If a number is divisible by 10, then the number is even. **14.** If you want to learn about computers, then try our Computer Wizard tutorial for thirty days. **15.** You want to learn about computers; try our Computer Wizard tutorial for thirty days.

Reteaching with Practice

1. "It is raining outside" is the hypothesis. "There are clouds in the sky" is the conclusion.

2. "Two angles are vertical angles" is the hypothesis. "The two angles are congruent" is the conclusion. **3.** If a number is even, then it is divisible by two. **4.** If two angles are congruent, then the measures of the angles are equal.

5. Argument 2 is correct. The hypothesis (it is noon on Monday) is true, which implies that the conclusion (the children are in school) is true by the Law of Detachment. **6.** If I throw the stick, then my dog will bring it back to me.

Lesson 2.6

Warm-Up Exercises

1. $x = 9$ **2.** $x = -23$ **3.** $x = 29$ **4.** $x = 9$ **5.** $x = -4$

Daily Homework Quiz

1. hypothesis: the temperature begins falling; conclusion: the rain will change to snow.

2. If an angle is a right angle, then its measure is 90°. **3.** The two lines never meet. **4.** If the rain becomes sleet, then school will be delayed.

Practice A

1. D **2.** E **3.** C **4.** F **5.** A **6.** B

7. Subtraction Property of Equality

8. Multiplication Property of Equality

9. Addition Property of Equality **10.** Division Property of Equality **11.** Substitution Property of Equality **12.** Substitution; Subtraction

13. Transitive; Symmetric

14. Reflexive Property of Equality

15. Symmetric Property of Equality

Practice B

1. $m\angle A$ **2.** GH; EF **3.** $m\angle 1$; $m\angle 3$ **4.** $\overline{KL}$

5. $\angle 6$; $\angle 5$ **6.** $\overline{AB}$; $\overline{EF}$ **7.** Subtraction Property of Equality **8.** Multiplication Property of Equality **9.** Addition Property of Equality

10. Division Property of Equality

11. Substitution Property of Equality

12. Midpoint; Segment Addition; Substitution; Division or Multiplication. **13.** Substitution Property of Equality; Subtraction Property of Equality

Copyright © McDougal Littell Inc.
All rights reserved.

Lesson 2.6 *continued*

Reteaching with Practice

1. Transitive Property of Congruence
2. Symmetric Property of Congruence
3. Angle bisector; Angle bisector; Transitive, Congruence **4.** $x = 3$

Quiz 2

1. 70° **2.** 110° **3.** $y = 20$ **4.** If <u>two angles are vertical angles</u>, then (the angles are congruent.)

5. If <u>it is sunny</u>, then (I will wear my sunglasses.)

6. If it is summer, then Halie will go swimming.

7. Transitive Property of Equality.

8. Symmetric Property of Congruence.

9. Reflexive Property of Congruence.

10. Division Property of Equality.

Review and Assessment

Chapter Review Games and Activities

[1]B		[2]C	O	N	C	L	U	S	I	O	[3]N
I		O									E
[4]S	Y	M	M	E	[5]T	R	I	C			V
E		P			H						E
C		[6]L	I	N	E	A	R	P	A	[7]I	R
T		E			O					F	
S		M			R					T	
	[8]S	E	G	M	E	N	[9]T			H	
		N			M		R			E	
		T					U			N	
		[10]A	D	J	A	C	E	N	[11]T		
		R							W		
[12]R	A	Y			[13]C	O	M	M	O	N	

Test A

1. 13, 13 **2.** 20.5, 41 **3.** $(-3, -1)$
4. $(3, -1)$ **5.** 39°; 39° **6.** 49°; 98° **7.** $x = 22$
8. $y = 18$ **9.** Complementary; nonadjacent
10. Supplementary; adjacent **11.** 79°; 169°
12. 33°; 123° **13.** 130°; 50°; 130°
14. 65°; 115°; 65°

15. If <u>it rains</u>, then (you will need an umbrella.)

16. If <u>an angle is a straight angle</u>, then (its measure is 180°.)

17. The angles are supplementary. **18.** We win the game. **19.** If the perimeter of a square is 24 inches, then the area of the <u>square</u> is 36 square inches. **20.** $m\angle D$ **21.** $\overline{PQ}$, $\overline{TU}$

Test B

1. 17, 34 **2.** 5.7, 11.4 **3.** (4, 1) **4.** (3, 10)
5. $(1, -1)$ **6.** 11°; 22° **7.** 60.5°; 60.5°
8. $x = 16$ **9.** $x = 10$ **10.** 33° **11.** 80.5°
12. 143° **13.** 65.5° **14.** $x = 10$; 130°; 50°
15. $x = 28$; 124°; 56° **16.** 133° **17.** 99.5°
18. 106 **19.** 120 **20.** 134

21. If <u>tomatoes are ripe</u>, then (they are red.)

22. If <u>it is raining</u>, then (you will need an umbrella.)

23. If you wash your uniform in bleach, then you will have to buy a new one.

24. Angle Addition Postulate **25.** Angle Addition Postulate **26.** Addition Property of Equality **27.** Substitution Property of Equality

SAT/ACT Chapter Test

1. C **2.** B **3.** A **4.** B **5.** A **6.** A **7.** C
8. D **9.** D

Alternative Assessment

1. Complete answers should include a clear and concise explanation of the Midpoint Formula, an example (done correctly) using the Midpoint Formula, and a diagram clearly showing the endpoints and midpoint of a segment.

2. a. $\overline{DH}$ **b.** $\overrightarrow{AF}$ **c.** *Sample answer:* $\angle DAC$ and $\angle CAB$, $\angle DAE$ and $\angle EAF$ **d.** *Sample answer:* $\angle FAC$ and $\angle CAB$ **e.** $\angle GAB$

f. $m\angle 1 = 58°$, $m\angle 2 = 58°$, $m\angle 3 = 32°$, $m\angle 4 = 32°$, $m\angle 5 = 58°$, $m\angle 6 = 90°$

g. complement; Given; Substitution; Subtraction

Copyright © McDougal Littell Inc.
All rights reserved.

Review and Assessment *continued*

Project: Origami

1.–2. Check that the origami figure matches the construction of the diagram. **3.** Check answers.

Sample descriptions of folds:

Valley fold: to fold the paper in front along the dashed line.

Mountain (or peak) fold: to fold the paper behind along the dashed line.

Reverse fold: to fold the paper so that two or more layers are contained within a crease.

Cumulative Review

1. A **2.** *Sample answer:* $\overleftrightarrow{EF}$, $\overleftrightarrow{EC}$ **3.** *A, E, F* and *G*

4. *Sample sketch:*

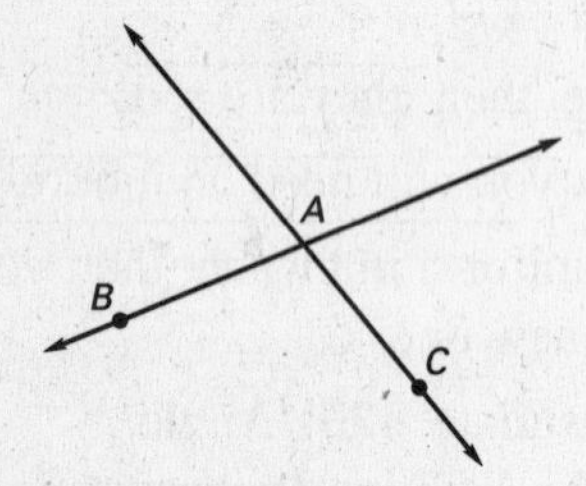

5. *Sample sketch:*

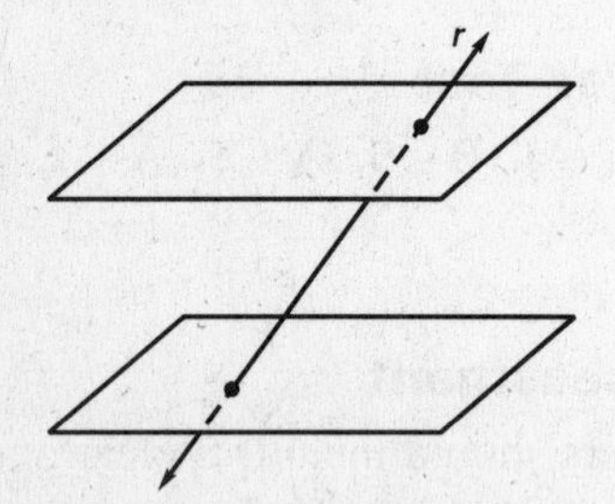

6. 6 **7.** 5 **8.** straight **9.** obtuse **10.** acute **11.** right **12.** (10, −3) **13.** (−3, 1) **14.** 17° **15.** 12° **16.** 114° **17.** $x = 5$ **18.** $x = 7$ **19.** 128° **20.** 53° **21.** 53° **22.** 55° **23.** 72° **24.** Any two of these pairs: $\angle AGB$ and $\angle EGD$; $\angle BGF$ and $\angle CGE$; $\angle AGE$ and $\angle BGD$; $\angle FGD$ and $\angle AGC$; $\angle BGC$ and $\angle FGE$; $\angle CGD$ and $\angle AGF$ **25.** $x = 45$ **26.** Your floors will get wet. **27.** Symmetric Property of Equality

28. Reflexive Property of Congruence

29. Transitive Property of Equality

30. Substitution Property of Equality

Copyright © McDougal Littell Inc.
All rights reserved.